ThermoPoetics

ThermoPoetics

Energy in Victorian Literature and Science

Barri J. Gold

The MIT Press
Cambridge, Massachusetts
London, England

For information about special quantity discounts, please e-mail special _sales@mitpress.mit.edu

This book was set in Stone serif by Toppan Best-set Premedia Limited. Printed and bound in the United States of America.

Library of Congress Cataloging-in-Publication Data

Gold, Barri J., 1966–
Thermopoetics : energy in Victorian literature and science / Barri J. Gold.
   p.   cm.
Includes bibliographical references and index.
ISBN 978-0-262-01372-7 (hardcover : alk. paper)
1. English literature—19th century—History and criticism. 2. Physics in literature. 3. Literature and science—Great Britain—History—19th century. I. Title.
PR468.S34G65   2010
820'.9'356—dc22

2009019766

10 9 8 7 6 5 4 3 2 1

I am grateful to the Modern Language Association for permission to reprint my article, "The Consolation of Physics: Tennyson's Thermo-dynamic Solution" (*PMLA* 117, no. 3 [May 2002]:449–464), which is revised and included as chapter 1 of this book.

For Andrew and Laney and Kendall

Energy and persistence conquer all things.

—Benjamin Franklin

Contents

Acknowledgments   xi

I The Consolation of Physics: Discovery   1

Prologue: Physics for Poets   3

Introduction: That Thing We Do   13

1 Tennyson's Thermodynamic Solution   33

II Energy and Empire: Applications   69

2 Grand Unified Theories, or Who's Got GUTs?   71

3 The Reign of Force   113

4 A Far Better Rest: Equilibrium and Entropy in *A Tale of Two Cities*   151

III The Engine and the Demon: Transformations   185

5 *Bleak House*: The Novel as Engine   187

6 Bodies in Heat: Demons, Women, and Emergent Order   225

Epilogue   259

Appendix: "Report on Tait's Lecture on Force:—B.A., 1876" by
James Clerk Maxwell   263
Notes   265
References   307
Index   321

Acknowledgments

Writing a book is hard. Writing a first book while raising two small-but-fabulous children and teaching full time would have been impossible if it weren't for the help of others. In addition to Muhlenberg College, which has repeatedly provided financial support, I'd like to express special thanks to the many individuals who contributed their energies to this project: To Heidi Reicher and Francesca Coppa for their truly energetic input and unfailing enthusiasm. To Russel Kauffman for our inspiring collaborations and the steadfast interest that always helped keep the ideas flowing. To Jefferson Pooley and Jeremy Teissere for the discussions that kept me on task and up to date. To Steve Lewis and Charlotte Zales for their painstaking attention to the final drafts. To my mom, Estelle "Teddy" Gold, who not only did so much with the kids, but also proofread the whole. And more love and thanks than I can express to my spouse, Andrew, for his unceasing and expansive support and for never letting me forget that it really does matter.

1   The Consolation of Physics: Discovery

Prologue: Physics for Poets

Energy is eternal delight.
—William Blake

$E = mc^2$
—Albert Einstein

Through a nice piece of luck, I happened to be at the International Centre for Theoretical Physics in Trieste, Italy, in January 2001, where a bumper sticker reading "Conserve Energy" had been affixed to an office door. In spite of the obviously good intentions behind the message, someone could not resist scrawling on it, "The Fools! Don't they know energy conserves itself?" This exchange attests not only to the sense of humor enjoyed by physicists the world over, but also to the ambiguity that haunts the "conservation of energy." Now where there is ambiguity, the literary among us will not be far behind. Whatever our feelings toward physics, whether or not these are deeply rooted in adolescent trauma, we are personally and professionally okay with words. Suggestive, problematic, sticky, contentious, culturally constructed, laden with baggage, protean, evolving, delicious words. If you happen to be more at home with irony, ambiguity, and metaphor than with force, mass, and acceleration, you will be pleased to know that what we now call physics, especially nineteenth-century physics, can be done—was in fact often conducted—in words. Indeed, much of the work of developing energy physics has been the work of wrestling with words. Old words had to be rethought and newly applied. Meanings had to be thrashed out. And occasionally, when nothing else would serve, a new word would be coined, including the word *physicist* itself.[1]

The rest of this prologue is an introduction to or a review of some of the basic terms and concepts of energy physics—conducted exclusively in words. Read it, skim it, skip it—everything will be redefined as needed anyway—or return to it when you feel you need a refresher. In what follows, I turn to the ways a little bit of physics can change how we look at Victorian literature. Meanwhile, I want to share my conviction that physics is a lot more like poetry than one might think.

Energy

Somehow, between William Blake and Albert Einstein, *energy* took on new meaning. Isaac Newton's disdain for the notion had made energy a faux pas in physics through the early nineteenth century. It was a metaphor, a word to describe people, a pathetic fallacy, a word predominantly for poets. Obviously, something has changed. Certainly, no contemporary physicist is likely to find "E = Eternal Delight" a particularly useful conversion factor. Very specific meanings, mathematical definitions, nasty-but-useful things like Hamiltonian operators, have emerged to delineate what *energy* means to the scientific community. And yet, it is a word still on everybody's tongue.

As *energy* began, in the mid-nineteenth century, to wend its way back into the good graces of science, it did so by building on a well-established reputation of social and metaphorical usage. So much so, that two Victorian writers can begin an account of the new science with the assertion that "Energy in the social world is well understood."[2] They proceed, by way of social metaphor, to introduce us to *kinetic energy* (energy of things in motion) and *potential energy* (energy stored or available for use). That they can do so suggests that by 1868, much of the transition in the meaning and status of *energy* that would be effected by the Victorians, had been effected. As the word and concept of energy emerge from the consolidation of disparate observables, things that look different—most notably heat, light, electricity, magnetism, gravitational attraction, and mechanical work—come to be understood as manifestations of the same thing. For a while, some will call it *force* and refer to the conversion and conservation of forces. Eventually, we will pretty much settle on *energy*.

The Laws of Thermodynamics

Before diving into its laws, we may as well have a stab at *thermodynamics*. The term, coined by the physicist William Thomson in 1854 at a meeting of the Royal Society of Edinburgh,[3] refers to that branch of physics that deals with the relations between heat and *other forms of energy*. When Thomson introduced it, he limited the latter to electricity and mechanical action. The word *thermodynamics* itself evokes the roots of the science of energy in the study of the movements, flow, interactions, and generation of heat. And while I will often use the term interchangeably with the somewhat anachronistic "energy physics" or the more culturally specific "science of energy,"[4] one always refers to the two fundamental principles of this science as the "laws of thermodynamics." In my opinion, these should be read with the lilting tones of a short Dickinson poem, in which the second clause invests the first with irony, poignancy, and maybe even a bit of self-abnegating humor:

1. The energy of the universe is constant.
2. The entropy of the universe tends towards a maximum.[5]

This simple statement of the laws of thermodynamics was originally made by another physicist, Rudolph Clausius, in 1865 and can be productively read as a good news/bad news joke. The good news is that energy is always conserved. The first law states that though energy may change forms, it can be neither created nor destroyed. For some, this was particularly consoling because it reinforced the notion that such creation or destruction belonged to God alone, that it was and always would be beyond the capacities or hubris of humanity. Isaac Asimov, taking a less godly perspective, explains that "the law of conservation of energy tells us we can't get something for nothing, but we refuse to believe it."[6] For most, however, the bad news—the "odious twin"[7]—is the second law: Entropy is always increasing. Or as Woody Allen put it, "It's the Second Law of Thermodynamics: Sooner or later everything turns to shit."[8] The upshot is this: while the energy of the universe does remain constant, that by no means ensures our continued access to it.

The Conservation of Energy

According to the first law of thermodynamics, the energy of the universe can't, in fact, run out. And it doesn't require any human intervention for its conservation. In rejecting the phrase "conservation of energy," the Victorian philosopher Herbert Spencer raises exactly this objection—that the word *conservation* implies "a conserver and an act of conserving,"[9] neither of which is actually necessary. He also objects to the term *energy*, preferring "persistence of force" to sum up the principle in question. But his objections anticipate the chuckles of a twenty-first-century wise guy in Trieste, as well as the more serious ways in which the language of energy resonates. No fools, we know perfectly well that energy conserves itself. And yet, we take very seriously the exhortation to conserve energy. So what do we mean when we say *conserve*? We can't protect energy from running out, and we don't need to. We must mean something about how well we use energy as we convert it from one form to another. In fact, we delineate usable energy from energy broadly speaking. That is, of course, what we mean when we refer colloquially to energy. We can, and really should I think, make efforts to conserve usable energy (or more exactly, sources of usable energy), which does not conserve itself but runs out or rather transforms—much to the dismay of the Victorians as to our own society—distressingly quickly.

Entropy: The Poet's Choice

In a "comical chemical history" detailed in Charles Dickens's weekly magazine *Household Words*, young Harry Wilkinson attempts to educate his family on the rudiments of thermodynamics.[10] Taking "the flame of the candle" as his "little shining case," Harry asks,

"What becomes of the candle . . . as it burns away? where does it go?"

"Nowhere," said his mamma, "I should think. It burns to nothing."

"Oh, dear no!" said Harry, "everything—everybody goes somewhere."[11]

Harry's faithful account of the candle's fate (and our own) asks and answers a fundamental question about energy. As the candle—the source of usable energy—runs out, where does the energy go? Since the laws of thermodynamics obtain always and only within a closed system (including, we presume for these purposes, the universe itself), it doesn't actually leave. It doesn't go away. It can't disappear. Rather it changes, "goes to" as we sometimes say, other forms. One of these may be mechanical effect or work. Another and generally less desirable form is heat. As we try to articulate where we are in the progress of these transformations, which always result in less usable energy than we had previously, we discover or invent the concept of entropy. Clausius introduced this term as well, choosing it in a rather poetic impulse to find something that resonates with energy, "intentionally form[ing] the word *entropy* so as to be as similar as possible to the word *energy*."[12] Entropy is thought about differently by physicists than by chemists. It can be defined mathematically, and has even been usefully connected to a similarly defined mathematical quantity known as *information*. But defining *entropy*, like entropy itself, can be decidedly messy. Less a thing itself than a way of thinking about the state of things, entropy is roughly synonymous with *disorder*. Of course, if we want clear-cut definitions, disorder leaves us hardly any better off than we were with entropy. On the other hand, if we value what the poet John Keats called "Negative Capability" and prefer to spend our days dwelling in "uncertainties, Mysteries, doubts,"[13] we can't do better than to vacation in a universe of entropy.

That said, it is nonetheless helpful to have along a relatively simple explanation of the concept. A basic engineering defini- tion of entropy associates it with the bit of waste we must make every time we try to accomplish any work. We can also think of entropy as a measure of energy beyond our use. Of course, understanding entropy as a measure of energy we can no longer use, cannot help but resonate (to our deeply ingrained new-historicist, deconstructionist, postcolonial sensibilities) with social implications: What use? And who's *we*?

Diffusion, Disorder, and the Paradox of Heat

Momentarily finessing the question of who *we* are, we users of energy like it in nice, tight, localized forms that we dub energy sources. The mechanical potential energy at the top of a water-fall, the chemical energy in fossil fuels, the nuclear energy contained within the matter of a nearby star, "the white energy of boiling water," and the "full-grown energies of heaven" all constitute energy sources.[14] We think of systems that have usable energy nicely contained in its place—hot spots separate from cold spots, high position from low—as orderly. Thermodynamically speaking, discernible *difference* constitutes order. The second law guarantees, however, that this kind of order can never be more than a temporary state of affairs: "There is a tendency in the universe to change the superior kinds of energy into the inferior or degraded kinds."[15] Energy in its place tends to spread out or diffuse. Eventually it all transforms irrecoverably into the random motion of molecules, into heat at uniform temperature, slipping away into the vastness of global and universal heat sinks. There is an immense amount of thermal energy diffused throughout the cold of the world's oceans, even more in the cosmic background (black-body) radiation of space. But, as if on a permanent bad-hair day, we can't do a thing with it. And when all of the energy has gone over to the dark side, we call that the *heat death* of the universe—a highly entropic state of affairs in which the usable energy has all been used up. Thus the tension that plagues the laws of thermodynamics comes to rest with its most familiar term: *heat*. Technically defined as "thermal energy transfer," heat more broadly construed becomes its own opposite, referring both to certain forms of usable energy and to their untimely end. Heat may allude to heat sources—hot spots, clearly delineated from cold spots, which we can use to run an engine—or it may refer to the uniformly lukewarm state of affairs when all such usable, localized heat has diffused.

Engine Design, or How I Learned to Stop Worrying and Love the Second Law

The heat death of the universe is enough to dampen the spirits of even the most incurable optimist. (Hence, *Bleak House*.) The good news is, it's a long way off. In the meantime, we may still occupy ourselves with the more pressing question of how to make the best use of what we have. As energy makes its inexorable way toward maximum entropy, we can get in the way. Usable energy will transform to unusable forms, regardless of whether we use it and how. But we have used it to great effect before now. Arguably, the imperative of the second law is responsible for life itself. The second law has recently been reframed as the principle that "nature abhors a gradient." Life, then, can be understood as deriving, benefiting from, and accelerating universal processes of gradient reduction.[16] Not so ambitious as to take the thermodynamics of life as my model, I prefer to think in terms of the engine. Energy degrades; an engine, like life, gets in the way. Broadly speaking, an engine is anything that uses energy. More specifically, an engine uses energy to produce *work* or *mechanical effect*, to do something we find useful, like exert a force over a distance (a phrasing that suggests at once the equation that defines work and its potentially imperial resonances).

Regrettably, an engine must speed the increase of entropy, negotiating the "tension between a short-term release of energy and the long-term price paid for that release."[17] Considering the trade worthwhile, Victorian engineers built better engines, hoping to get that much more work done before all the energy is sunk into heat. It isn't everything, but it's enough to ignite an industrial revolution, enough to reshape the novel, enough for the qualified thermodynamic optimism that drives *ThermoPoetics*.

Introduction: That Thing We Do

Physics is like sex; sure, it may give some practical results, but that's not why we do it.

—Richard Feynman

The French philosopher Michel Serres has called J. M. W. Turner "the first true genius in thermodynamics."[1] That sounds great. But what on earth does it mean to call a Romantic landscape painter, with decidedly impressionistic tendencies, a genius in thermodynamics, the science that deals with the relationships between heat and other forms of energy? In addition to engaging in some rather promiscuous disciplinary mixing, Serres's claim also has a small problem of chronology, for the French engineer Sadi Carnot's essay "On the Motive Power of Heat" was written in 1824—twenty years before Turner makes one "see matter." But apparently unmoved by the priority claims of engineers or scientists, Serres insists on the precedence of Turner's public vision:

From [the artist George] Garrard to Turner, the path is very simple. It is the same path that runs from [the mathematician Joseph] Lagrange to [the physicist Sadi] Carnot, from simple mechanics to steam engines, from mechanics to thermodynamics—by way of the Industrial

Revolution. [Before this change] wind and water were tamed in diagrams. One simply needed to know geometry or to know how to draw. Matter was dominated by form. With fire, everything changes, even water and wind. Look at *The Forge*, painted by Joseph Wright in 1772. Water, the paddlewheel, the hammer, weights, strictly and geometrically drawn, still triumph over the ingot in fusion. But the time approaches when victory changes camps. Turner no longer looks from the outside; he enters into Wright's ingot, he enters into the boiler, the furnace, the firebox. He sees matter transformed by fire. This is the new matter of the world at work, where geometry is limited. Everything is overturned. Matter and color triumph over line, geometry, and form. No, Turner is not a pre-impressionist. He is a realist, a proper realist. He makes one see matter in 1844, as Garrard made one see forms and forces in 1784. And he is the first to see it, the very first. No one had really perceived it before, neither scientist nor philosopher, and Carnot had not yet been read. Who understood it? Those who worked with fire and Turner—Turner or the introduction of fiery matter into culture. The first true genius in thermodynamics.[2]

Serres describes the onset of a new worldview. Left behind is a Newtonian, Lagrangian, mechanistic vision dominated by "forms and forces." We now see the forge through the lenses of hot and cold, work and energy. Casually reconceiving impressionism as he goes, Serres thus paves the way for those of us who love science almost as much as we love chatting about art and literature. He articulates what we always secretly believed but didn't know to whom to confess it: that art, literature, and science work *together* to form and reform how we understand the world. And that sometimes art and literature may even come first. Without agonizing over who was the first or what it means to be a *true* genius, I would nonetheless venture that Tennyson was *a* genius in thermodynamics and that Dickens was a damn good engineer.[3] On top of that, the physicist James Clerk Maxwell was undoubtedly a poet, and if his equations of electricity and magnetism are rather more elegant and timeless than his occasional verse, the two together suggest how analogous, intertwined, and mutually productive, poetry and physics may be.

Such claims also suggest a certain studied disregard for priority. I have tried hard to resist what has been called the "diffusion model" of science and society[4]—manifest in the dominant assumption that science may influence literature, but not the other way around. It is, in fact, quite clear to me that literature has often, perhaps always, influenced science, especially in the delicate, early stages of a scientific development, before a phenomenon has been named or a hypothesis adequately articulated. Literature participates in creating as well as expressing the cultural milieu in which science happens.

For instance, consider the poem *In Memoriam*, finally published in 1851 by Alfred, Lord Tennyson, to commemorate a friend who had died almost twenty years earlier. *In Memoriam* expresses not only Tennyson's personal grief, but also many of the anxieties wrought by early nineteenth-century science, especially the fear of extinction brought on by the findings of paleontology, the worry that "God and Nature [are] at strife" and that nature itself is wasteful and cruel. Having worked through these anxieties—the emotional or *affective* effects of the science— in ways that earn him the title of Poet Laureate (a position he held for over forty years), Tennyson also helps set up Charles Darwin for the remarkable reception accorded to the *Origin of Species* in 1859. By that time, Tennyson has already developed a notion of evolution as onward-and-upward progress that brings us individually and as a species nearer to perfection and even to God. So when Darwin writes, even in his first edition, that "all corporeal and mental endowments will tend to progress towards perfection,"[5] his audience is primed to experience this as mitigating the more grim implications of his work.

And when James Clerk Maxwell clearly alludes to *In Memoriam* in a letter to the natural philosopher Michael Faraday, saying that Faraday's "lines of force can 'weave a web across the sky,' "[6] Maxwell suggests how literary metaphor may help to shape scientific ideas. The usefulness of metaphor in scientific thinking is even more emphatic within Maxwell's poem "Reflections from Various Surfaces" (1853), which explores the likeness of light and water, nicely anticipating the fluid analogy through which Maxwell eventually develops his equations of electricity and magnetism. And that such usage is part and parcel of science more broadly is clear, for example, in the musings of the physicist and brewer James Prescott Joule, when in 1847 he announces his belief in what will become the law of conservation of energy. The coining of the term *kinetic energy* is still two years away, so Joule falls back on a seventeenth-century term, "*vis viva*, or *living force*," identifying the term as metaphorical "inasmuch as there is no life, properly speaking," even as he emphasizes that "it is *useful*."[7]

Productive Conversation

Throughout this work, I will demonstrate my belief that science and literature engage in a mutually productive conversation. The action of specific metaphors, like Tennyson's "a web is woven across the sky," can be seen in the development of Faraday's "lines of force" into field theory. Such a metaphor thus helps to enable the conceptual jump from an idea like Newton's, that forces can act at a distance (as when one body attracts a far-away body through gravitation), to a notion that objects change the character of space around them, creating fields, which in turn act on other bodies. Bigger metaphors that shape a variety of narratives circulating through a culture serve both the development and the dissemination of scientific ideas. Thus one scholar can identify entropy as a "root metaphor" and another can account for the rapid spread of anxiety associated with the second law of thermodynamics, because this development in physics participated in an already widespread mythology regarding the death of the sun.[8]

Moreover, the methods of science and literature overlap. Ursula Le Guin suggests, in an introduction to her novel *The Left Hand of Darkness*, that science fiction deploys a new set of metaphors to think about the as-yet unarticulated: "The novelist says in words what cannot be said in words." Similarly, Darwin's *Origin of Species* may be taken to struggle with metaphor in order to use "words . . . thus paradoxically."[9] His is an exemplary case of the development of scientific ideas through analogy, metaphor, and personification. Postponing the explication of natural selection until chapter 4, Darwin first discusses at length its human analogue in artificial selection, or what we more often call *breeding*. His discussion suggests strongly that the metaphor of nature-as-selector is not *merely* a mode of expression, a way to explain a new idea, but essential to the development of the idea itself. As he revises the *Origin*, he interjects such concessions as "if I may be allowed to personify" and here and there adds the word *metaphorically* into his original "it may be said."[10] In doing so, he may shift his emphasis, but the text retains traces of the original functions of his metaphor(s), functions that enable thinking as much as articulating.[11] As we consider the development of energy physics, we will see that literary and scientific methods overlap. I have suggested above a few of the ways metaphors function in physics at this same moment, as in Maxwell's thinking about light-as-fluid, but my first extended example of literary method in physics will suggest that such overlap is not exclusive to metaphor, because early energy physics partakes strongly of the structure and function of elegy.

Of course, literature and science also converse when science is deployed *as* metaphor, especially when already-established scientific terms and concepts find their way into poetry and fiction to serve the exploration of some more manifestly anthropocentric concern. This most familiar form of the exchange is not without its own lingering puzzles. I have made an effort to be interested in but not oppressed by the questions that are a continuing challenge to the study of literature and science: "What do the parallels signify? How do you explain their existence? What mechanisms do you postulate to account for them?"[12] In the last thirty years or so of scholarship, we have seen that many mechanisms account for the conversation between literature and science. Some mechanisms are easy to trace, as when Maxwell evokes Tennyson, or Tennyson is known to have read Robert Chambers, Humphry Davy, and Charles Lyell. In a similarly emphatic move within *Bleak House*, Dickens's man of industry names his son "Watt." Sometimes, it's harder to see the connections. Herbert Spencer was certainly close to the physicist John Tyndall, but did the novelist Edward Bulwer-Lytton ever actually meet or read Spencer? Or are all the striking parallels we will explore in chapter 2 *just* coincidence? Certainly they coincide, but I have to take issue with the *just*. If the connection is not simply causal, the correlation is nonetheless compelling. And if we cannot (do not even want to) ascribe some simple model of influence—preferring instead, for example, to consider how both literature and science draw on "root metaphors" or perpetuate developing mythologies, deploy similar methods of reasoning, and address the same cultural concerns—we may nonetheless find the study of the parallels highly suggestive.

We are, moreover, very lucky that interdisciplinarity is in vogue at the moment. Studies of literature and science have been (back) around for long enough to see that such a conversation exists. It is an increasingly small population willing to assert that science or literature develops in isolation or that either is an endeavor that does not partake of and contribute to the larger cultural milieu. Unfortunately, interdisciplinarity is not so easy to do or to do well. True interdisciplinarity, I would suggest, requires that we cross disciplinary lines not only in our choice of subject, but also in our choice of audience. We need to be able to speak across disciplinary lines, from literature to physics, and back again. To historians and sociologists. To folks trained in Victorian studies and those trained in science studies, and indeed, to readers whose interests in energy or literature we cannot anticipate.

This book must operate at the intersection of many fields, and yet I know perfectly well that it cannot be all things to all people. Instead, I hope to offer something to anyone whose interests intersect with my own, no matter where those intersections might occur. I have tried to write *ThermoPoetics* to be as inclusive as I know how, to invite readers to find affinities and shared interests. In my effort to reach across disciplinary boundaries, I have refrained, as best as I could, from addressing any group of specialists to the exclusion of others. This means that I have had to forgo, for example, the kind of scholarly apparatus and indeed the kind of prose I would have used if I were writing exclusively to literary scholars. The chapter on Tennyson suggests the origins, but not the ends, of this book. Anyone wishing for a sense of the book I ultimately did not write is invited to

take a look at the version of this chapter published in the May
2002 issue of *PMLA*. In those early days, I saw myself as a scholar
of Victorian literature who happened to be interested in physics.
I attended carefully and almost exclusively to the needs and
wishes of a literary community. The prose is professionally
dense. Indeed, several friends in physics claim to have lingered
up to half an hour on the first paragraph alone. And while these
particular friends might be somewhat given to hyperbole, I took
their point. Similarly, the text of that article dedicates more
than a few paragraphs to summarizing relevant scholarly work
on the poem—something useful, interesting, necessary to scholars of Tennyson, the Victorian period, or literature more broadly,
but potentially off-putting to the differently educated reader. As
far as possible, I have tried to omit such paragraphs here, by
which I mean no disrespect to the many literary scholars from
whose work this book benefits tremendously. In many cases, I
am fully conscious of the debt I owe. No doubt, in many cases,
I am not; for when a literary idea (like a scientific fact) has really
made it, it tends to obscure its own origins, seeming to diffuse
without any human intervention. I offer my thanks to my intellectual benefactors—known and unknown alike. And where
these are known, and their gifts closely tied to the concerns of
this book, I offer further thanks in the form of endnotes.

More extensive discussion of current scholarship is eschewed
in the interests of conservation of scholarly energy. For I do, in
fact, believe that we can't get something for nothing. And I
believe the benefit has been worth the cost. It is my hope that
readers will find the insights valuable even where the methods
and mechanisms are unfamiliar. As I have done. For while I am
trained neither as a historian, nor as a sociologist of science, I
have learned a great deal from what some of them have had to

say. Thus, while I have not, for example, taken Bruno Latour's advice and followed scientists as they go about their work, I have nonetheless found his suggestions for doing so tremendously helpful in thinking about the discursive links between literature and science. And because I am greatly indebted to the work of many scholars, across these fields and in the loosely bound area of science studies where they meet, I would invite readers to benefit similarly.

And so, I ask that you be willing to think along with me about the connections between things we tend to dissociate. I imagine that you are comfortable with and perhaps even interested in literature, if not necessarily familiar with these particular texts. You may be comfortable with physics, or you may be profoundly uncomfortable for the moment. If the former, I hope that Enrico Fermi was right when he said, "Never underestimate the joy people derive from hearing something they already know."[13] If the latter, it is my intention that, by the end of this book, you will be versed in the terms and principles of energy physics, well beyond ordinary cocktail party standards. Similarly, I would invite you to have a look—if you have not already done so—at some ideas about what makes a scientific fact and how they get around and indeed, how literature may participate in their making as well as their circulation.

Cultural Splitting

This is, perhaps, a lot to ask, and so in writing this book, I have tried hard to be the friendly, excitable, profoundly unintimidating person I am in *real life*. (One more caveat, and then I will be finished.) I apologize in advance for any excess of silliness that may result, but as vices go, I prefer that to its opposite—too serious a tone, too highfalutin language, especially when the latter is defensive. Because in spite of the present vogue for interdisciplinarity, I have found the messages to be mixed to the point of hostility. Between college and graduate school, I went from an interesting oddity among physicists to an oddity—interesting at best, scary at worst—among literature scholars. At MIT, I could write, and some of my classmates good-naturedly suggested packing me off to Harvard. In graduate school at the University of Chicago—where many of my literature friends but not one among the physicists had to pay his or her own way—the teasing was less good-natured. I tucked my interest in physics quietly away with all information regarding my financial package. Years later, I have come out of the closet in the relative safety of a small liberal arts college in Allentown, Pennsylvania. At Muhlenberg College, I am allowed to teach such courses as "Literature and Science" and "Energy, Elegy & Empire"; I may, in fact, even use the word *science* in my course descriptions (though not the word *physics*). Here I have found a small inter-disciplinary community of similarly interested individuals as well as the resources to keep in touch with a larger community, especially through the networks of the Society for Literature, Science, and the Arts, the Interdisciplinary Nineteenth-Century Studies community, and the British Society for Literature and Science. And yet, so many of us still exclude ourselves from this promising pursuit on the mistaken premise that our present discomfort with or disinterest in physics is irremediable and permanent—or worse, a necessary correlative to a true love of literature.

This experience is hardly new or isolated. In a 1959 lecture titled "The Two Cultures and the Scientific Revolution," C. P. Snow recounts his dismay at meeting many highly educated people who have expressed "with considerable gusto . . . their incredulity at the illiteracy of scientists. Once or twice," he recounts, "I have been provoked and have asked the company how many of them could describe the Second Law of Thermodynamics. The response was cold: it was also negative." The "coldness" of the response is more than just an opportunity for thermodynamic punning. C. P. Snow's famous identification of "Two Cultures" is a mid-twentieth-century response to mostly Victorian changes in the status of scientists and methods of education. It marks at once defensiveness, discomfort, and hostility, all of which respond to an implicit shifting—well underway by 1959—in what Snow calls the "standards of traditional culture."[14] That we so often construe *scientific literacy* as opposing more traditional notions of literacy is an unfortunate and relatively new development.

And the problem with such cultural splitting is not just academic. With the rising energy and environmental crises, it is more important than ever that we talk to each other, and more important that we all understand enough about science to participate in the public discourse that I hope will get us out of this mess. We are, after all, in many ways like the Victorians—only more so. In fact, our best scientific efforts predict a much more imminent doom than theirs ever did. While Victorian estimates for the death of the sun placed the event at least thousands and often billions of years away, our scientists predict that we will run out of fossil fuels in roughly 30 to 100 years at our current rate of consumption. (The difference depends largely on whether we are willing to revert to coal and risk the increased production of greenhouse gases.)[15] Why then are we so much less freaked-out than the Victorians seem to have been? We ought properly to be feeling less optimistic now than even William Thomson was when he predicted with "certainty, that the inhabitants of the earth cannot continue to enjoy the light and heat essential to their life, for many million years longer," for it seems even less likely to us that "sources now unknown to us are prepared in the great storehouse of creation."[16] Our relative complacency may derive in part from the fact that we are less in the habit of talking to each other than the Victorians were. Despite the pride we take in having created an information society, we still have not managed to tap into what history teaches us about how science operates in culture and vice versa. Certainly, I would want every person—at the very least, every educated person—to know enough about energy to make good choices at the polls. And it would not be a bad thing to learn something about the affective content of scientific discourse, as well as the coping mechanisms that allowed our predecessors, and would allow us, to maintain a qualified and educated optimism as we pursue solutions to some very big problems.

Energy's Brighter Side

Now, if I may lighten the tone a bit, I would add that such a conversation is bound to pay off not only in the long-term benefits to our culture and species, but also in the more immediate production of literary pleasure—something I take very seriously. I believe that talking about physics *and* literature effectively doubles our pleasure. And so I have chosen to treat the confluence of energy physics and texts we already love, including Tennyson's *In Memoriam*, Dickens's *A Tale of Two Cities* and *Bleak House*, Stoker's *Dracula*, as well as the work of slightly less familiar Victorians like Herbert Spencer and Edward Bulwer-Lytton. Is this choice of texts, as one scholar worries, "capricious"?[17] Well, yes and no. In spite of various whimsical elements in their treatment, not one was chosen *merely* on a whim. They were chosen because we love to read them, teach them, turn them into miniseries. And of course, they were chosen because they work. Each literary text is so fully suffused with a thermodynamic sensibility that it's a wonder to me that any of my chapters actually end. On the other hand, I make no claims to have covered the vast number of Victorian texts productive of and informed by the discourse on energy. Quite the contrary; I'd be overjoyed if you would go forth and find lots more. After all, oddity loves company.

If you should decide to engage in the (thermo)dynamic reading practice I am proposing, *ThermoPoetics* is not a bad place to start. The readings of Tennyson, Dickens, Stoker, Wilde, and so on, ground a decidedly literary introduction to each new element of thermodynamic thought: conservation and dissipation, the linguistic tension between force and energy, the disillusionment of equilibrium, the paradox of heat, the quest for a grand unified theory, the practical mechanics of heat engines, the strategies for coping in an inexorably entropic universe, and the demonic potential of the thermodynamically savvy individual. I will show that literary texts embrace the language and ideas of energy physics to address the familiar concerns that dominate the Victorian textual fabric—about religion, evolution, race, class, empire, gender, sexuality, decadence, as well as the shape of the narrative itself. These concerns, in turn, shape the hopes and fears expressed by and about the new physics. *ThermoPoetics* thus serves at once to focus a new lens on Victorian literature and to provide in-depth examples of its practical applications. And in order to delineate a modest thermodynamic canon, this book treats key but accessible essays by noted scientists (Thomson, Joule, Carnot, Stewart, Lockyer, Tait, and Tyndall) and even Maxwell's poetry—enough to provide the grounding for readers inspired to take up the energetic treatment of Victorian literature.

The choice of the Victorian moment is fortuitous as well, for not only did the Victorians face a prospect of energy depletion not unlike our own, but they also enjoyed particularly cordial relations between physics and literature. Before the splitting of the so-called two cultures of the sciences and the humanities, well before the "science wars" of the 1990s, and after the height of Romantic resistance to what Blake calls "Newton's sleep,"[18] nineteenth-century literature and science engaged frequently with each other as if the connections were natural, obvious, useful. This is not to say that relations were not subject to tension as well as analysis. But poets, novelists, and physicists—when not worrying about whether they should—shared ideas and methods and assumptions as they went about their different but overlapping work. They were, after all, educated together. For most of the nineteenth century, university education offered rather limited choices (theology, classics, mathematics, or, after 1828, English literature[19]). Darwin studied theology. Maxwell wrote poetry as he studied mathematics at Cambridge, and "many of the early signs point[ed] to a flowering in literature rather than science."[20] And the correspondence between such university-educated, mostly wealthy, mostly male/ Victorians continued after they left university to pursue the specific interests for which we now mostly remember them. While I cannot applaud the exclusivity of this conversation— the fact that it was dominated by white Anglican men of a certain class—I can nonetheless appreciate the effects of its inclusivity, the apparently natural exchange between the literarily and scientifically inclined members of society.

Victorian energy physics, moreover, may have had an added incentive to keep up the correspondence with literature, for it was deeply engaged in pondering imponderables. In retrospect, we can see why scientists found it hard to think about what were called imponderables—such as heat, light, electricity, and magnetism—once they had given up thinking of them as liquids (albeit rather elusive liquids that could not be weighed). Literary methods, such as metaphor, which make it possible to say in words "what cannot be said in words,"[21] must have seemed more than usually handy. And since many scientists had what we now would call literary aspirations and acted as their own popularizers, speaking not only at the British Association but also at their local church, publishing not only in *Nature* but also in *Blackwood's*, the *Fortnightly Review*, and the *Encyclopaedia Britannica*, the conversation did indeed reach a broad, generally educated audience.

Is it possible, then, that the relations between literature and physics in this moment were in fact so cordial, compared to what came before and after, that we rarely notice them? The complex relationship between the physical sciences and literature in the Victorian era has, to a surprising degree, been overlooked by literary scholars. While historians of science, writing about the experiences of scientists, have repeatedly commented on the way energy physics played out in the broader culture (including an elite literary culture), scholars working on Victorian *literature* have paid relatively little attention to some very powerful connections. Indeed, historians have produced several "unashamedly cultural histor[ies]," which attend not only to the development of scientific ideas, but also to their connections to the larger cultural milieu.[22] Except, however, for occasional gestures toward Mary Shelley, H. G. Wells, or the poetry of James Clerk Maxwell, these publications have strayed

little from the dominant concerns of historians—the workings
of what we loosely term the "real world." But historians themselves are converging with literary scholars in their assessment
of the importance of text, even fictional and poetic text, in
understanding history.[23] And every teacher of undergraduates
knows that a fictional text can illuminate patterns of belief
and expectation that are less obvious in other forms. Here
we encounter the general milieu, the fears, doubts, frustrations,
fantasies, hopes, dreams, and aspirations of the historical
moment.

And yet, there has never been a book-length treatment of the
conversation between Victorian literature and thermodynamics.
There have been excellent collections of essays treating Victorian literature and science.[24] Energy physics, however, remains
conspicuously absent or subordinated to other concerns (such
as glaciation or Tyndall's reputation). Several prominent scholars of Victorian literature and science have touched on aspects
of Victorian thermodynamics,[25] and Bruce Clarke's *Energy Forms*
dwells on late Victorian literature as he theorizes the relationship between entropy and allegory, paying particular attention
to energy and economics before moving to reconfigurations of
entropy among the modernists. Among the classics of Victorian
literature and science, George Levine's *Darwin and the Novelists*
clearly articulates some of the conventional wisdom about thermodynamics that *ThermoPoetics* seeks to reverse—for example,
the assumption of a "conflict between . . . the progressive vision
of Darwinism and the degenerative vision of thermodynamics."
And Gillian Beer devotes a lovely chapter of *Open Fields* to the
most powerful cultural anxiety advanced by Victorian thermodynamics: "The Death of the Sun." Here, Beer also observes "the
apparent contradiction between the 'progressive' implications

of evolutionary theory and the emphasis in the physics of Helmholtz and Thomson on the ageing of the sun."[26]

This very brief sampling hints at some of the contributions I hope to make to this ongoing conversation. One is to shift away from the impulse to over-Darwinize; a second is to reassess the emotive content of Victorian scientific discourse, what I will call "scientific affect." Without wanting to detract from Darwin's brilliance as a writer and naturalist, to derogate the importance of his ideas to Victorian culture and vice versa, or to underrate the importance of evolution to contemporary biology, I would nonetheless like to help create a more balanced perspective. The discourse on energy is at least as prevalent as Darwinian biology in Victorian sensibilities, and perhaps even more important to us. Why we tend to focus on evolutionary theory to the exclusion of thermodynamics is unclear. Undoubtedly, both evolutionists and physicists were engaged in struggles for cultural authority. The Victorians may have felt compelled to choose, since there was a highly publicized conflict between the predictions of thermodynamics and the premises of evolution. William Thomson, in particular, wanted it made clear that the estimations of the duration of the sun's heat, past and present, simply did not allow enough time for natural selection to be feasible. But evolutionary and thermodynamic narratives participated in the same larger cultural discussion, evinced similar anxieties, and drew on similar methods to assuage these anxieties. For us, this means that the treatment of Victorian literature and science could do with some reassessment and redistribution. Texts that seem to scream *Darwin* also reveal their deep investment in thermodynamic concerns. These texts include those by Darwin himself. For example, his discussion of rudiments in *The Descent of Man* exhibits some of the same preoccupation with waste

reflected in the practical concerns that gave rise to thermodynamics. He rethinks the value of organs that are apparently of
little or no use to an organism, such as the nipples on males.
These, he claims, are the *most useful* for scientific interpretation,
since "useless and rudimentary organs [prove] by far the most
serviceable for classification."[27] In this, Darwin's efforts strongly
echo the efforts of engineers to reduce the waste of usable
energy evident in even the most efficient engines.

As we will see in chapter 1, Tennyson registers and reconciles
similar anxieties about the wastefulness of nature in ways that
suggest how a narrative of evolution-as-progress goes hand in
hand with the capacity—so important in Tennyson—to replace
second-law anxieties of inevitable loss with a firm first-law belief
in conservation and tranformation. Tennyson's case, in turn,
suggests that evolution is not necessarily as positive a narrative
nor thermodynamics as negative as either might have seemed.
We then can see what Edison knew to great personal advantage:
that energy has a bright side. Throughout *ThermoPoetics*, I
have attempted to articulate a qualified thermodynamic optimism that runs throughout Victorian literary and scientific
productions.

In chapter 2, I move to the quest for a "grand unified theory,"
a desire to find one theory that will explain all natural and
indeed all cultural phenomena. For both Herbert Spencer's *First
Principles* and Bulwer-Lytton's *The Coming Race*, the unifying
theory seems to be energy. Both texts articulate evolutionary
and social narratives that are particular manifestations of the
larger, linked principles of conservation and dissipation. The
impulse to apply these principles to social as well as physical
systems, however, requires Spencer and Bulwer-Lytton to revisit
and revise their fantasies of equilibrium. Certainly Spencer's

phrase "survival of the fittest" and his enduring association with social Darwinism don't resonate with ideas about equilibrium. And for Bulwer-Lytton, the impossibility of sustained equilibrium eventually lead to one profound social implication: that imperial expansion is physically inevitable. But in chapter 3, I explore other social possibilities implicit in thermopoetic discourse—even the suggestion that the "reign of force" is over—especially as these appear in popular accounts of Victorian energy physics and in the poetry of James Clerk Maxwell.

Chapters 4 and 5 explore two novels by Charles Dickens: *A Tale of Two Cities* and *Bleak House*. Dickens himself proves an adept novelistic engineer as he works within the constraints of thermodynamics to produce both order (in *A Tale of Two Cities*) and work (in *Bleak House*) in spite of the universal drive to entropy. In doing so, Dickens manages to transform entropy itself. To some the herald of universal decay, entropy becomes in his hands a mere complication (however intractable) in the building of better engines.

The final chapter then explores the productive potential of the entropically savvy individual. In chapter 6, *Dorian Gray* and *Dracula* are read "against the grain" such that anxieties about the Other (the foreign, the female, the demonic) reveal a double-edged narrative about the creative potential of chaos. This creative use of entropy is, moreover, one of many ways Victorian literature anticipates twentieth-century physics, not only non-equilibrium thermodynamics, but also chaos theory. With the epilogue, the book returns to its roots, with the hope that the conversation between literature and science will continue to be ongoing and mutually productive, their relationship long, complex, and fruitful.

1 Tennyson's Thermodynamic Solution

HANNAH: Is there anything in it?

VALENTINE: In what? We are all doomed? (Casually.) Oh yes, sure—it's called the second law of thermodynamics.

HANNAH: Was it known about?

VALENTINE: By poets and lunatics from time immemorial.

—Tom Stoppard[1]

Could Thomasina Coverly, a thirteen-year-old girl living in a country house in Derbyshire, possibly have discovered the second law of thermodynamics in the spring of 1809, over four decades before the term *thermodynamics* was even coined? In fiction, where such things happen, she can and does. And though this rather remarkable teenager exists only as a character in Tom Stoppard's 1993 play *Arcadia*, her unlikely discoveries trigger the very real suggestion that in 1809 we knew. In some sense, we have always known. Byron knew in 1816, when he opened the poem "Darkness" with these lines:

I had a dream, which was not all a dream.
The bright sun was extinguished, and the stars
Did wander darkling in the eternal space,
Rayless, and pathless, and the icy earth
Swung blind and blackening in the moonless air.[2]

And in the same sense, Alfred, Lord Tennyson knew in 1851, when he finally published *In Memoriam*, a really long, important Victorian elegy that he had begun writing almost two decades earlier on the death of his friend Arthur Henry Hallam. By the time it was finished, *In Memoriam* had greatly extended its reach, not only commemorating a lost friend, but also addressing many of the concerns and anxieties, hopes and aspirations, characteristic of early Victorian England. Within its hundred-or-so pages, coping with the loss of a friend becomes thoroughly entangled with coping with other kinds of loss: a threat to faith; a sense of instability wrought by changes in the economy and social structure; anxieties associated with advances in technology and science, with landscape, foreign relations, the map of Europe, gender relations, conceptions of race, and education, and of course, with the loss of nature's usable energies soon to be associated with the second law. The poem spoke to many; the consolation it effected was widespread. Queen Victoria was said to have kept it at her bedside, and Tennyson was made poet laureate shortly after its publication. *In Memoriam* was undoubtedly what so many elegies are not; it was a crossover hit.

The poem's name alone suggests that it is engaged in remembering that which might otherwise be lost; as a memorial, its job is a kind of conservation of memory. Now savvy readers are already anticipating a parallel that this chapter pursues at length, as we extend *In Memoriam*'s reach just a bit further. For it speaks to us, too—enabling us to recover a moment prior to what one might call *thermodynamic memory*, when notions of conservation and dissipation are finally inscribed in and as the laws of thermodynamics. Tennyson thus seems to fulfill the promise of the poet and the lunatic—not to mention the genius. He seems to know things before the rest of us. Or perhaps more accurately, he seems to have said in words what we could not yet say in words. Some might even say that within his poem, we find ideas in circulation, before they are properly commodified.

In this chapter, I want to look at *In Memoriam* in a way that is both familiar and new. It is familiar, because we have spent a century and a half considering this poem as somehow representative of the spirit of the age, the zeitgeist, the cultural milieu or political unconscious. It is new because *In Memoriam* has not been properly appreciated as the brilliant work of thermodynamics that it almost certainly is. As such, it allows us to study the literary production of ideas we think of as scientific. *In Memoriam* suggests, if nothing so simple as the influence of poetry on science, at least something more conversational—something I have come to call *ThermoPoetics*, a term intended to suggest mutual influence, common concerns, and even simultaneous discovery. And it gives us a sense of how such a thermopoetic conversation might go, how poetry speaks to science, how it says in words what cannot yet be said in words, how tender, young, unformed scientific ideas reflect and inform the other concerns that pervade what we will loosely call a culture, and even how a culture prepares itself to embrace such ideas.

Useful Anachronism

Alfred, Lord Tennyson's *In Memoriam* is a strange mix of past and future. On the one hand, Tennyson embraces certain distinctively Romantic beliefs, especially the notion that poetry is a way of knowing. At the same time, he embraces a characteristically Victorian investment in scientific modes of inquiry. Tennyson thus insists on the consonance of the "two cultures" of science and literature at a moment when they seem to be diverging. And *In Memoriam* not only evinces a deep concern with the science of the day, but also disrupts expected patterns of influence. Even where readers have sought to explore the interaction between science and literature, they tend to posit a unidirectional influence, considering almost exclusively the influence of science on literature. Why we do this relates to what Bruno Latour has called a "diffusion model" of how scientific facts move about. Studying science "in action," Latour does much to mess with our comfortable assumptions that these have a life of their own, that somehow they exist and circulate spontaneously, without any human help. This "diffusion model" tends to operate more as an unexamined assumption than as an articulated model and implies that scientific facts don't need any help from one distinctly human endeavor—poetry.[3]

In Memoriam suggests otherwise.[4] It not only circulates such facts, but arguably even helps them to exist. The first clue comes from Tennyson's apparent prescience in scientific matters— something readers have noted for over 150 years. We are by now thoroughly familiar with *In Memoriam*'s struggles with the anxieties wrought by evolutionary narrative. Published eight years prior to Darwin's *Origin of Species*, its phrase "Nature, red in tooth and claw" has been "vested by historians with the power to sum up nothing less than the impact of evolutionary thought on Christian humanism."[5] In fact, some of "the sections of 'In Memoriam' about Evolution had been read by his friends some years before the publication of the *Vestiges of Creation*,"[6] which Tennyson requested from his bookseller, after seeing an ad suggesting that the work "contain[ed] many speculations with which [he had] been familiar for years, and on which [he had] written more than one poem."[7] It seems that *In Memoriam* anticipates not only Darwinian evolution, but also one of its significant precursors in early nineteenth-century science.

But that's not all. The 1850s witness the consolidation of two sciences; both evolutionary biology and energy physics emerge in this moment as viable—indeed, as what one famous scholar calls "dominant paradigms."[8] But while Tennyson's engagement with biological and geological thought is well known, his conversation with the physical sciences has been largely overlooked.[9] Tennyson's own, oft-quoted to-do list—"Monday. History, German./Tuesday. Chemistry, German./Wednesday. Botany, German./Thursday. Electricity, German./Friday. Animal Physiology, German./Saturday. Mechanics./Sunday. Theology"[10] —coupled with the posthumous testimony of his friend, the astronomer Norman Lockyer, suggest an early and abiding interest in the physical sciences. Indeed, Thursday's regimen as well as Saturday's suggest that in spite of Romantic antipathy to certain Newtonian ways of knowing, there is a place for physics in poetry.

However, my concern here, as I have suggested, is less with what Tennyson knew than with what he could not possibly have known: the laws of thermodynamics. We are hard-pressed to pinpoint when these laws were first articulated. Indeed, for one historian, "the hypothesis of energy conservation . . . publicly announced by four widely scattered European scientists" between 1842 and 1847, is an exemplary case of simultaneous discovery.[11] Some locate the genesis of energy physics considerably earlier: Thomas Young's 1803 lecture "On Collision" seems to offer one of the earliest usages of the term *energy* in its modern physical sense.[12] But Young focuses on the motion of macroscopic objects. He identifies what will come to be called kinetic energy, which is only one subcategory of what energy will come to comprehend once the Victorians get their hands on it. A French engineer by the name of Sadi Carnot also gets a lot of credit for inventing thermodynamic theory. His 1824 paper "On the Motive Power of Fire" is often taken to be the earliest formulation of what I have been calling the "law of entropy" or the "second law"—the law that posits the inevitable loss of useful energy. William Thomson will later call this work "a perfectly clear and general statement of the 'Conservation of Energy'"—that is, of the first law of thermodynamics.[13] (Thomson, if you recall, has already shown up in this book as the one who coined the term *thermodynamics*. His work was so valued that he was eventually knighted, becoming Baron Kelvin, a name that will probably be more familiar to scientists and engineers, who undoubtedly recall the absolute temperature scale he developed, in which measurements are made in degrees Kelvin.) Anyway, Thomson was largely responsible for the circulation of Carnot's essay in English. And his attention suggested that Carnot's contribution was undoubtedly significant. Still, it isn't exactly clear what that contribution was. Was it about conservation or dissipation? How could it have been both? And even if we think we understand what Carnot's contribution was, even if we understand it largely as having to do with conservation, it seemed to be contradicted by the work of James Prescott Joule, the Manchester brewer whose experiments showed heat *loss* far more than he seemed to like. Thomson sought to reconcile this increasingly disturbing contradiction in his essay "On the Dynamical Theory of Heat" (1851) (see Thomson 1882). The considerable conceptual work this required suggests that the scattered pronouncements of a principle here, a definition there, did not suffice to establish thermodynamics as a science or, more precisely, a physical theory.[14] It was not until 1854 that Thomson identified this science as "thermodynamics"; not until 1865 that Clausius coined the key term *entropy*.[15] A very broad popularization of thermodynamic theory followed, marked not least by John Tyndall's *Heat: A Mode of Motion* (1863) and by Balfour Stewart's *The Conservation of Energy* (1873). Thus, much of the development of thermodynamic theory coincided with—and virtually all of its popularization followed—the writing and publication of *In Memoriam*.

Nonetheless, *In Memoriam* is saturated with the language of energy physics. Though "energies" itself appears only twice,[16] the concepts that *energy* eventually comprehends—heat, light, power, force—surface again and again, as do images that suggest the concerns of thermodynamics more broadly: loss and gain, waste, systems, the behavior of gases, order and disorder, and changes of state or form. In Tennyson, as in the emergent science of energy physics, these terms evince considerable overlap and—like the things they represent—they tend to transform into one another. As James Prescott Joule will observe in 1847, "All three, therefore—namely, heat, living force, and attraction through space (to which I might also add *light* . . .)— are mutually convertible into one another. In these conversions, nothing is ever lost."[17] Similarly, in Tennyson, light and heat, life and attraction, will prove interconvertible and will be governed by the same principles. And when read through the lens of constructive anachronism—the same kind of vision that enables us to identify retroactively the significance of contributions such as Young's and Carnot's—the poem's repeated, apparently disconnected returns to these concerns emerge as a coherent thermodynamic narrative.

Drawing on the same culture of science in and through which the physicists developed their ideas, reading Laplace and Kant, among others,[18] Tennyson can be said to have discovered— poetically—not only the terms, but also the principles and processes of the nascent science of energy physics, especially the poetic evocation of the tension between conservation and dissipation that haunts the first and second laws of thermodynamics. Of course, saying such a thing requires us to stop and rethink what we mean by *discovery*. Usually, when we think of discovery, we imagine that one of us found (uncovered, recognized, introduced) something that was already out there, in something we call nature. Then, all we have to do is give it a little push and it diffuses—so to speak—on its own; it's so compelling, so true, so simply factual that except for a stubborn few, people just accept it.[19] I'd like to use the term *discovery* somewhat differently to include not only the circulation, but also the shaping of what we come to know as scientific fact. In this kind of discovery, physicists as well as poets participate in a process by which facts are made—not out of nothing, but made nonetheless. Poetry, I would suggest, offers one of many "strategies that give the object the contours that will provide assent,"[20] making something (a fact, an idea, a physical theory, a scientific object) we can actually work with, think about, build on, accept or resist.

In this expanded sense, then, *In Memoriam* exemplifies poetic discovery simultaneous with—even prior to—scientific discovery. As such, the poem requires us to reexamine our expectations of the relation between Victorian poetry (or religion for that matter) and science, especially physics. Far from being antagonistic or mutually exclusive endeavors, poetry and science draw on the same language and in many ways wrestle with the same contradictions as each develops the principles physics will call the laws of thermodynamics. Working out how Tennyson anticipates these laws, we also elicit the ways he reshapes what seem like familiar tropes of Romantic elegy. As he deploys these

tropes in the context of his own scientific concerns, they resonate in important physical, as well as spiritual, ways. At the
same time, *In Memoriam*'s conversation with energy physics
leads us to revisit our ideas about relations in and among Victorian science: we find elegiac echoes within the discourse of
energy physics itself. And we discover that Victorian physics
and biology may have enjoyed an affective relation quite counter
to what we have come to expect.

A Brief Scientific Interlude: The Laws of Thermodynamics and the Paradox of *Heat*

The *affective relation* between physics and biology refers to the
emotional weight attached to each as they come to permeate
popular conversation. By the second half of the nineteenth
century—significantly, in the wake of the publication of *In
Memoriam*—evolutionary theory seemed relatively optimistic.
Victorians invested the term *evolution* with the promise of progress (a promise, biologists insist, by no means inherent in the
theory); it was taken to imply an onward and upward development—of individual, species, race, and nation—into increasingly perfect forms. So much so, in fact, that when evolution
worked in what was considered a downward or regressive direction, the process was often given a different name: *devolution*.
Thermodynamics, on the other hand, was widely experienced
as the scientific basis for universal pessimism; it seemed to
promise only decay, dissipation, degradation, and death. The
second law in particular "seemed from the start to run counter
to the optimistic 'progressivist' directions of most contemporary
science, particularly evolution."[21] This affective opposition,
moreover, found reinforcement in a professional one: William

Thomson and his followers were the most vocal scientific opponents of Darwinism. Indeed, Thomson himself remarked on the impossibility of Darwinian evolution within the universal time scales allowed by thermodynamic theory.[22]

In Memoriam, however, is not constrained by late-century expectations of thermodynamic pessimism. So I would suggest we try to revisit, as best we can, both the poem and the laws of thermodynamics without the pressure of late Victorian affect. The simple statement of these laws originally articulated by Rudolph Clausius in 1865 is one that we still use (the annotations, in italics, are mine):

1. The energy of the universe is constant. *The first law implies that energy can be neither created nor destroyed. In a closed system, though energy may change forms, the total energy is always conserved.*

2. The entropy of the universe tends towards a maximum. *"Entropy" is the term given to the measure of disorder in a system. The second law thus implies that in a closed system, energy always changes to increasingly less orderly, less usable forms.*[23]

The first law, the "conservation of energy," seems to promise that nothing can be lost. It operates in affective opposition to the second law, which in threatening the perpetual and irreversible increase of entropy, suggests that everything must be lost. Nonetheless, the combination suggests not only a tension, but also a careful balance, an elegant parallelism. Indeed, Clausius "intentionally formed the word *entropy* so as to be as similar as possible to the word *energy*."[24] Both laws, moreover, imply the necessity of closure; failing other modes of closure, the universe itself acts as the ultimate closed system. And both are implicitly, though centrally, concerned with change—the *dynamics* in *thermodynamics*. Indeed, identifying, articulating, and resolving the apparent contradiction between loss and conservation was nothing short of the work of integrating important but loosely connected observations into the science of energy physics (more on this later), but we may understand the physical resolution in simple terms as follows: Yes, energy is conserved. It can take many forms, including heat and mechanical work. But once it has been transformed into heat (more precisely, into heat at a uniform temperature), no work can be done with it. Thus energy is conserved, but it becomes unavailable for use.

Heat itself is a tricky term, which carries some of the tension we see between the two laws. In physics at least, this term generally represents energy in its least useful, most entropic or diffuse state. *Heat death* then refers to a state of things in which all energy has been, not lost, but transformed, albeit irrecoverably and uniformly, into heat. Of course, in popular parlance, heat death tends to be associated not with the excess, but with the loss of heat. Heat is its own opposite; its popular usage suggests usable energy, what scientists and engineers call *heat sources*, bodies at higher temperatures from which we can derive warmth or run steam engines. However, as we have seen, *heat* in technical parlance just as often signals energy in its least usable form; here the word evokes the heat sink, those "waste places" that form the repositories of energy past its usefulness.[25] Thus *heat* itself proves a "contradiction on the tongue"[26] that reproduces linguistically the tension between the first and second laws.

In short, while there is no fundamental physical contradiction (a resolution that required a good deal of negotiation) between the first and second laws, thermodynamics is laden with tension. And the interplay between conservation and dissipation structures a central tension within Tennyson's poem, as within Victorian thought more broadly. But where (except for certain diehard optimists) the second law came to dominate a Victorian mindset increasingly concerned with dissipation and degradation outside of, as in conversation with, popularizations of thermodynamics, Tennyson's willed optimism shaped itself according to a first-law sensibility that rendered the second law not merely palatable, but hopeful. *In Memoriam*, as I will argue, though driven by loss, is finally able to find consolation by subordinating loss to the larger concept of change, to hold loss and conservation in tension, to effect a careful balance between the two. In this way, the intellectual work done by *In Memoriam* parallels that done by the founders of thermodynamics. Both are the work of negotiating contradiction, of holding apparent oppositions in well-balanced tension, of shaping the whole in a way that people would not only accept, but even find consolingly consistent with certain widely held and cherished beliefs about how the world works. Section 1 (completed about 1834) lays out this consolatory program, suggesting—in terms that look uncannily thermodynamic—that the poem's work will be "to find in loss a gain to match," to soothe second-law anxieties through the promise of first-law compensation.

"Spring No More": Waste, Death, and the Second Law

William Thomson's 1862 announcement in *Macmillan's Magazine* of "the age of the sun's heat" triggered a widespread cultural anxiety that encompassed no less than the cooling of the world and the death of all things as the sun burned itself out. And while such an event may have been predicted "by poets and lunatics from time immemorial," the second law of thermodynamics brought new urgency and new form to this ancient fear. To a public just recovering from the fossil-induced anxiety of extinction that Tennyson articulates so nicely—"From scarpéd cliff and quarried stone/She [Nature] cries, 'A thousand types are gone'"[27]—the death of the sun seemed not just inevitable, but frightfully imminent. Depictions of the sun's death, newly energized by scientific authority, "pass[ed] rapidly into uncontrolled and mythologized form."[28] By way of illustration, we can look to a fiction that was particularly troubled by the sun's imminent demise. In H. G. Wells's *The Time Machine* (1894), the unnamed time traveler "watch[es] with a strange fascination the sun grow larger and duller in the westward sky, and the life of the old earth ebb away."[29] Wells captures the fear that William Thomson famously expresses, "that inhabitants of the earth cannot continue to enjoy the light and heat essential to their life, for many million years longer."[30]

Some time earlier, Tennyson also depicts a dying sun. "The stars," he writes, "blindly run;/A web is woven across the sky,/ From out waste places comes a cry,/And murmurs from the dying sun."[31] Tennyson is evoking Laplace's Nebular Hypothesis, but his lines also strongly suggest the second law. Though Laplace's theory postulates a mechanism for solar origins (through the cooling and contraction of a cloud of gas or nebula that formed the sun and planets), Tennyson here focuses on its implications for endings. The death of the sun is linked to an anxiety about the blindness and cruelty of nature evident in the poem's evolutionary narrative, even as it evokes a broader cosmological concern through the image of waste space. The term *waste*, which has not yet given way, in physical theory, to *dissipation*, *disorder*, or especially *entropy*, also suggests a further Victorian anxiety that would attach to the second law. For added to the increasing conviction that the sun was limited as a power supply was the realization that most of the sun's energy would be wasted: how little of the sun's heat and light (late

Victorians calculated anxiously) would be intercepted for use on earth; how much would dissipate uselessly into space! One more thing: this moment is also an interesting (and traceable) example of this poem's contribution to scientific thought. In writing to Faraday about whether he can transfer his methods for modeling the behavior of electrical interactions to a model of gravitation, Maxwell formulates this possibility in words that are self-consciously evocative of Tennyson: "then your lines of force can 'weave a web across the sky.'"[32]

The general waning of power, moreover—suggested above by the transformation of a "cry" into mere "murmurs"—returns emphatically in a later image of heat death: "I dream'd there would be Spring no more./That Nature's ancient power was lost;/The streets were black with smoke and frost."[33] This moment exemplifies how Tennyson operates at the intersection of Romantic poetry and Victorian physical theory. The end of the universe as that which occurs in dreams echoes those lines from Byron's "Darkness" quoted earlier: "I had a dream. . . ." But in Tennyson, the ultimate loss of power, the end of heat and light, is that which occurs in nature, as well as in dreams. In this way, it is more like the scientifically driven fictions of H. G. Wells or of Camille Flammarion, author of the 1894 novel *Omega: The Last Days of the World*. The poem's concern with nature failing, at least implicitly scientific by this time, confirms the anxieties *In Memoriam* repeatedly expresses: Tennyson's "Nature" almost always evokes the anxieties wrought by science, most especially the threat of science to faith (as in "Are God and Nature then at strife?"[34]). And when *She* does finally give out (yes, Tennyson's "Nature" is quite explicitly female), it is specifically her "power" that is lost. This word choice is significant in contemporary as in Victorian physical theory; alongside *force*, *work*, and *energy*, *power* has a thermodynamic ring. Indeed, at this moment, *energy* has yet to take its central place in thermodynamic language. "Motive power" (*puissance motrice*), on the other hand, is the term Carnot, for one, uses in the 1824 essay that is so evocative for those who build on his work.

But as you've heard before, that's not all! It is not only Tennyson's use of language that places him squarely in conversation with emerging physical theory. His impulse to locate analogous phenomena at vastly different scales suggests how his poetic methods dovetail with those of his contemporaries in physics. The same concern with the loss of heat that here colors Tennyson's cosmos applies to the death of the individual and to the metaphorical death of day with nightfall. This mode of analogy is characteristic of thermodynamic discourse; physiologists were among the earliest advocates of energy theory, and by century's end, the depiction of the body as a thermodynamic system or of nightfall or an eclipse as a mini heat death will become familiar, if not commonplace.[35] Indeed, this mode of thinking is what—for one prominent Victorian physicist—distinguishes a physical theory, such as energetics, from an abstract science, such as Newtonian mechanics. Grounded in the observation of a wide range of phenomena, its principles must be "reduced to the form of a science"; in turn, this reduction is "the better the more extensive the range of phenomena whose laws it serves to deduce."[36] Eventually, Tennyson's friend, the physicist John Tyndall, will claim for thermodynamics a "wider grasp and more radical significance" than Darwinian evolution, which becomes merely one manifestation of the concept of energy.[37]

Tennyson too applies his principles to a wide range of systems. Personal death echoes cosmological death. Loss is tied to waste

on the personal scale, for where space itself is figured as entropic—a "waste place"—so is the end that the poet anticipates for
himself: "Somewhere in the waste," he says, "the shadow sits
and waits for me." In the absence of faith in an afterlife, it would
seem that "earth is darkness at the core/And dust and ashes all
that is."[38] Waste and entropy seem fundamental and final. Or
as two popularizers of the new science will soon put it: "The
principle of degradation is at work throughout the universe. . . .
As far as we are able to judge, the life of the universe will come
to an end not less certainly, but only more slowly, than the life
of him who pens these lines or of those who read them."[39] And
repeatedly throughout *In Memoriam*, the speaker's own death is
figured as the loss of heat, light, and even electricity. Thus he
worries, "How dwarf'd a growth of cold and night,/How blanch'd
with darkness must I grow!" and begs the spirit of his friend
to "be near me when my light is low." And like so many
things in thermodynamics, death itself transforms, manifesting
itself even as the loss of electricity, a time when "this electric
force, that keeps/A thousand pulses dancing, fail[s]."[40] Thus,
Tennyson not only evokes the physiological imagination that
animates Frankenstein's monster, but also links *electric* to *force*—
a term relatively well defined under Newton, but very much in
flux at this moment. Though a principle that finally cannot
hold in the face of emerging physical evidence, the conservation
of force (something Newton does not promise) is a key precursor
to the first law; such uses of *force* illustrate how it functions as
an early synonym of *energy*—though not the final word. Nor is
waste the last word in Tennyson; "dust and ashes" are not "all
that is." As I will argue, it is Tennyson's cognizance of what will
become the first law of thermodynamics, in many ways rooted
in Romanticism, that enables his famous consoling gesture on

both the personal and the popular scales—for the loss of his friend as for the rift between God and Nature, science and faith, produced by evolutionary and geological concerns.

Meaningful Metaphor: The Progress of *Energy*

If it weren't for the particular transformations they undergo, light, heat, electricity, and certainly death, might leave us comfortably enough in the realm of poetic tradition. After all, even *energy*, the keystone term of thermodynamic language, carried metaphorical and social weight before the physicists took it up. Let's briefly track these sources—so to speak—of *energy*: "Physicists had already borrowed the language and authority of social prophets," and indeed had borrowed the idea that the concept of entropy built itself on the foundation of an ancient commonplace of decline, irreversibility, and disorder.[41] Energy concepts, moreover, suggest a connection between Romantic philosophy and Victorian natural science. Nineteenth-century positivism has been called the "true nineteenth-century successor to the romantics' efforts at totalization."[42] One historian traces in considerable detail the link between the "central tenet of Romantic philosophy that nature should be apprehended as a coherent and meaningful whole" and early nineteenth-century work to demonstrate the underlying unity of physical forces.[43] And another holds that the Romantic "doctrine of the essential unity of all forces in nature leads directly to the law of conservation of energy," further identifying the shift in thermodynamic affect from mid to late century: "The first law of thermodynamics (conservation of energy), inspired in part by the philosophy of romanticism, provided an organizing principle for the science of the realist period. Likewise the second law of thermodynamics (dissipation of energy), which arose from the technical analysis of steam engines, provided a *dis*organizing principle which turned out to be highly appropriate for the neoromantic period."[44] This important distinction is generally glossed over in literary treatments of thermodynamics—a shift from first- to second-law dominance, which correlates with the difference between early- and late-century depictions of cosmological burnout and the consolatory potential that, as I will argue, Tennyson is able to find in thermodynamic principles at mid century. And of course, Tennyson brings poetry into the fold.

It is, perhaps, not surprising that the desire for Romantic wholeness should run strong in a poet like Tennyson, seeking consolation for the loss of a friend and restoration of a shaken faith, or that he should resurrect Romantic uses of *energy* to achieve these ends. The concept of energy would have had particular appeal for Tennyson, because "the classical or pre-scientific energy concept . . . operated ambivalently between physical and spiritual registers."[45] And energy's specific potential for elegiac consolation can be traced to Romantic poetry. *In Memoriam* in many ways echoes Shelley's "Adonais"—a poem Tennyson particularly admires, though he later declares that "Shelley had no common sense."[46] Even more striking is the way Tennyson's closing "one God, one law, one element" evokes Coleridge's "Religious Musings": "one Mind, one omnipresent Mind,/Omnific."[47] As energy, after a long period of disrepute in science dating from Newton's *Principia*, "re-emerged . . . for reasons primarily metaphysical, and especially religious, rather than physical," Coleridge draws a connection between this one Mind and energy, "declar[ing] it to be 'Nature's essence, mind and energy!' subsequently confiding that 'tis God/Diffused through all, that doth make one whole."[48]

It is this religious character, resonant in Tennyson's "full-grown energies of heaven,"[49] that makes the Romantic use of energy attractive to Tennyson and that, at the same time, begins to mark the divergence between his deployment of energy and theirs. What distinguishes Tennyson's energy-in-elegy, placing him squarely in conversation with the emerging science of energy physics, are the place of faith and the place of figurative language in his poem as well as in the new science—especially in relation to knowledge of the physical world. Although Coleridge, "a Victorian doubter before his time," may well have appealed to Tennyson precisely because he too wrestled with the increasingly visible gap between the truths of science and those of religion,[50] "Coleridge realizes that to proclaim is not to prove, and that the *sine qua non* for a faith such as the one he wishes to articulate is an unquestioning belief in God strong enough to lay aside any nagging questions."[51] Tennyson, by contrast, begins with huge, nagging questions. And far from reacting against the certainty of science (as the Romantics were inclined to do), he will draw on science as he formulates a response to his explicit questioning of faith. Moreover, where Romantic energy is decidedly extraphysical (what would a physicist make of Blake's "Energy is Eternal Delight"?), *In Memoriam*'s energetics attempt to marry the physical to the spiritual, to imagine the "soul,/In all her motion one with law."[52] Similarly, where the Romantics make claims for separate, poetic knowledge—what one scholar calls "romanticism's . . . supreme privileging of the artist as prophet-deliverer of a moribund social order"[53]—Tennyson does not strive to dissociate his from other, especially scientific, ways of knowing. Though Tennyson resists any simple materialism, the knowledge he seeks is not knowledge of the extraphysical, but of the physical and extraphysical as ultimately inseparable.

This meeting of physical and spiritual correlates with the changing uses of figurative language in poetic as well as in scientific inquiry. Romantic uses of energy are subject to what the Victorian critic John Ruskin will later call the "pathetic fallacy"; indeed the term *energy* seemed to many "a covert attempt to humanize the object-world through a species of anthropomorphic projection."[54] Tennyson, on the other hand, while rejecting the poetic as a uniquely privileged means of knowing, nonetheless retains metaphorical, or analogical, thinking as a powerful tool for knowing the natural world. Similarly, though opponents of the new physics would continue to raise such objections to *energy*—opting, by the 1870s, for the term *force* to signal their belief that however useful, the concept is still merely a "logical fiction"[55]—analogy and metaphor were increasingly acceptable tools for thinking within science.[56] And even Joule, as early as 1847, was hardly troubled by the figurative baggage attached to the term "*vis viva*, or *living force*." That expression, he notes, "may be deemed by some inappropriate, inasmuch as there is no life, properly speaking, in question; but it is *useful*."[57]

The usefulness of energetic metaphor extends beyond what Joule suggests here, and several scholars have noted the connection between the structures of energetics and those of metaphor. The concept of entropy becomes a deeply embedded and widely used cultural metaphor.[58] It seems ready-made to do so because of the structure of energy transformation itself. After all, Rudolph Clausius—in coining the term *entropy* "after the Greek word 'transformation' . . . borrowing the root of the term 'trope'—the linguistic torsion that produces nonliteral uses"—suggests a continuity between the metamorphic capacity of language and that of matter itself. "In the name of [entropy,] energic and linguistic transformation became metaphors for each other."[59] Transformation—refiguration—characterizes both poetic and scientific ways of knowing, because its structures inhere in the natural world that both seek to know. Tennyson's use of language thus proves not merely (as "pathetic fallacy") or exclusively (as "supreme privileging"), but meaningfully metaphorical, as a real way of coming to know the physical world, and indeed, as deeply apropos, in mirroring the physical world he at once describes and investigates.

"Power in Darkness": The Consolation of the First Law

The phenomenon-qua-metaphor that grounds Tennyson's investigation is heat. While images of heat loss dominate the early sections of the poem, Tennyson turns increasingly to the question of whether heat can also generate and be generated. As the poem progresses, heat figures increasingly as source, as that which can provide light or even life, that which generates and warms: "life is not as idle ore,/But iron dug from central gloom,/And heated hot with burning fears." Even in absence, heat figures as usable energy. The poem imagines the absent Hallam, for instance, sitting among his family "a central warmth diffusing bliss." Indeed, even allusions to light, life, and heat that are not realized figure rather strangely as something akin to what physicists will call potential energy: The "unborn faces [of Hallam's children] shine/Beside the never-lighted fire."[60] Thus Tennyson develops contrasting notions of heat sources and heat sinks (what Carnot calls "the source and . . . the refrigerator"[61]), notions that Tennyson, like a good physical theorist, can then apply to a broad range of phenomena. As he returns to Laplace's Nebular Hypothesis, he shifts his attention to its implications for beginnings. Replacing "I dreamed there would be spring no more" with "They say,/The solid earth whereon we tread/In tracts of fluent heat began,"[62] the poem now suggests the process through which (Laplace theorized) our sun and planets were formed. Looking at heat from both sides, Tennyson illustrates a central dilemma in the resolution of thermodynamic principles, even as he moves toward an increasingly optimistic vision of what those principles imply.

Similarly, much of Tennyson's energetic language—especially the developing theme of light, which was previously marked by loss—attaches increasingly to the capacity to generate or change. For a time, darkness signals predominantly loss and lack as Tennyson worries about a time "when my light is low" and imagines himself "on the low dark verge of life." The poem, however, becomes increasingly sure that the loss of light or electricity need not imply the loss of power. Darkness now evinces a power of its own: "And Power was with him in the night/Which makes the darkness and the light,/And dwells not in the light alone." This development of faith out of doubt, a conventional feature of elegy, is described in distinctly thermodynamic terms. Clearly steeped in the spiritual concerns of Romantic energetics, this passage also suggests a physical revelation—that dark and light are both power, differently manifested. Conversely, where darkness may figure as source, light itself may figure as sink. Thus " 'Farewell! We *lose* ourselves in light.' "[63] Tennyson's most complex thinking on light in its various manifestations, however, occurs in his 1849 Prologue: "Our little systems have their day;/They have their day and cease to be:/They are but broken lights of thee." Resonating with the poem's physical concerns (as in "star and system rolling past"[64]), "our little systems" suggests not only our belief systems,[65] but also our physical systems, our all-too-temporary bodies. In a move worthy of Einstein, these prove material manifestations of light itself. "Broken lights," moreover, implies at once disorder and transformation; in life or in death, our existence proves an entropic manifestation of divine light.

In terms increasingly distinct from Romantic elegy, death too figures as change—physical change—rather than loss: "I wage not any feud with Death/For changes wrought on form and face." Tennyson thus shifts from second-law anxiety to first-law hope, even as his thermodynamic language proliferates: *form*, *state*, *process*, *power*, and even *diffusion*. One of the many ways change is wrought, death is merely one part of an "eternal process moving on,/From state to state the spirit walks."[66] But change is not a sufficient—though it is a necessary—condition of conservation:

But thou art turn'd to something strange
    And I have lost the links that bound
    Thy changes; here upon the ground,
No more partaker of thy change.[67]

Change here clearly still implies loss. What, then, enables the shift from "I have lost the links" to the principle of conservation that eventually governs even the mechanics of friendship, such that "the all-assuming [all-destroying] months and years/Can take no part away from this"?[68] How is the second-law threat of an all-assuming entropic decay subordinated to the first-law promise that no part can be taken away?

As the poem reimagines death-as-change in a manner more consoling, it reconciles the problems wrought by this dissipative model by reconceiving these changes as "bound" within a larger system:

What are thou then? I cannot guess;
> But tho' I seem in star and flower
> To feel thee some diffusive power,
I do not therefore love thee less.[69]

The links are not lost; Hallam's "diffusive power" can still be
felt. It is "bound" on the ground—in flower as well as in star—
and is marked by a love that is not less. In this way, Tennyson
even anticipates the notion of "bound energy" that refers to
energy beyond our reach. (Like the enormous amount of heat
energy within the world's oceans, "bound energy" is there, but
we can't do any work with it. I will talk more about free and
bound energy when I get to Dickens.)

Meanwhile, I want to point out once more how—when it
comes to Romantic poetry—*In Memoriam* proves once again the-
same-thing-only-different. Tennyson's phrasing above certainly
retains traces of William Wordsworth's "Ode on Intimations of
Immortality from Recollections of Early Childhood," but with
important differences; Tennyson, it seems, *can* bring back "the
splendour in the grass [and] glory in the flower"[70] through his
distinctive deployment of the notion of diffusion. Though profoundly implicated in the development of the second law (a precursor to which may readily be found in Fourier's diffusion
equation), diffusion nonetheless allows Tennyson to concep-
tualize transformation without loss. All he need do now is
imagine a system that, though it may (as in Clausius) encompass
the whole of the universe, is nonetheless closed. Where once
"Nature's ancient power was lost," Hallam's is conserved, and
Nature as well as Hallam persist in a threefold mixing with God:

My love involves the love before;
> My love is vaster passion now;
> Tho' mix'd with God and Nature thou,
I seem to love thee more and more.[71]

The critical shift from *waste* to *vast*ness—etymologically linked words sharing the Latin source *vastus*—marks a rethinking of the universe, not as waste space, but as a very large, closed system in which things are never actually lost, but merely diffused. In an 1876 address to the Royal Institution of Great Britain, the brilliant physicist James Clerk Maxwell would find a similarly consolatory metaphor, effecting a similar transition from waste to vastness, within his conception of the ether that fills all space: "The vast interplanetary and vast interstellar regions will no longer be regarded as waste places in the universe, which the creator has not seen fit to fill with the symbols of the manifold order of His kingdom. We shall find them to be already full of this wonderful medium; so full that no human power can remove it from the smallest portion of space, or produce the slightest flaw in its infinite continuity."[72] Tennyson's vastness goes even further, ultimately proving not only conservative, but also productive, as "star and system rolling past,/A soul shall draw from out the vast/And strike his being into bounds"[73]—a counterentropic development hardly conceivable in science until the advent of chaos theory.

The Consolation of Physics

Tennyson's development of thermodynamic concepts within and from a tradition of poetic elegy raises a complementary question: To what extent do we find elegiac traces in the development of physical theory? Indeed, there are striking analogies between the conceptual work done by Tennyson and that done by Victorian energy physicists. In addition to the prominent place of analogical reasoning in both, we can see the significant workings and reworkings of faith and faithlike convictions in the development of thermodynamic concepts among

physicists—processes strongly reminiscent of the religious consolation for which *In Memoriam* is so well known. All of these
reworkings, moreover, seem to circulate around the reconciliation of apparently contradictory uses of *heat*—a reconciliation
effected by the reconception of loss (and generation) as transformation, and accompanied, as in Tennyson, by the restoration
of faith.

Although the belief "that God [is] love indeed"[74] is not explicitly at issue in their scientific investigations, both Thomson and
Joule repeatedly evoke a Creator, distinguished by the unique
capacity to generate or annihilate matter. Joule's 1847 lecture,
"On Matter, Living Force and Heat" (presented at St. Ann's
Church Reading Room), is emphatic regarding the place of God
in scientific reasoning: "We might reason, *a priori*, that such
absolute destruction of living force cannot possibly take place,
because it is manifestly absurd to suppose that the powers with
which God has endowed matter can be destroyed any more than
they can be created by man's agency."[75] Much of Thomson's
language is less explicitly religious but undoubtedly faithlike in
its structure. While he cannot say with Carnot (regarding his
belief that any heat loss is precisely compensated for by an
equivalent gain) that "this fact has never been doubted," he
does observe that "the truth of this principle is considered as
axiomatic by Carnot" and that in spite of doubts raised by
Joule's work, "I shall refer to Carnot's fundamental principle . . . as if its truth were thoroughly established."[76] However,
where Carnot's confidence rests on his contention that heat is
a substance—called *caloric*—and therefore subject to conservation laws associated with matter, Thomson's belief has no such
backing. By mid century, the caloric view of heat is well on the
wane, and belief in the conservation of what would eventually

be comprehended within energy—heat, force, *vis viva*—must be sustained as faith, indeed, as faith under siege by increasingly visible evidence in nature.

Embryonic thermodynamic theory went through a phase that one very influential philosopher and historian of science would call "resistance"—during which even its eventual proponents didn't know what to make of it and, in many ways, didn't much like it. Thomas Kuhn identified such resistance as characteristic of periods that precede what he called "scientific revolutions." Contemporary historians of science would frame the phenomenon rather differently—saying, perhaps, that the fact of entropy had not yet been given "the contours that [would] provide assent."[77] Some scholars might borrow an analogy from psychology, suggesting that both Thomson and Joule seemed to go through a kind of denial stage before they were ready to move to acceptance. Thomson, it seems, wanted to adopt what he took to be Carnot's view of the conservation of heat, but the contradictory implications of Joule's work surface irrepressibly. Thomson included Joule's view first as a footnote: "This opinion seems to be nearly universally held by those who have written on the subject. A contrary opinion however has been advocated by Mr Joule of Manchester."[78] A similar observation, promoted from the footnotes to the main body of the text, appeared the following year in his "Account of Carnot's Theory": "The extremely important discoveries recently made by Mr Joule of Manchester, that heat is evolved in every part of a closed electric conductor, moving in the neighbourhood of a magnet, and that heat is *generated* by the friction of fluids in motion, seem to overturn the opinion that heat cannot be generated."[79] Significantly, however, the most troubling—and the most provocative—part of this discussion was again relegated to the margins:

When "thermal agency" is thus *spent* in conducting heat through a solid, what becomes of the mechanical effect which it might produce? Nothing can be *lost* in the operations of nature—no energy can be *destroyed*. What effect then is produced in place of the mechanical effect which is lost? A perfect theory of heat imperatively demands an answer to this question; yet no answer can be given in the present state of science.[80]

Why footnote such an elegant assertion of the conservation of energy—and such a clear imperative for the goals of energy physics? For one thing, it had become increasingly clear that this was a reassertion of faith, rather than an assertion of well-substantiated scientific principles. Even the assertion that "no energy can be destroyed" is deceptively simple—and deceptively assured—since it was not clear what energy comprehends. That Thomson still sought a "perfect theory of heat" suggests that the relation between heat and energy was as yet unsettled. And of course, this moment was emphatically about the absence of answers. It is noteworthy that though generation and loss threaten the principle of conservation in logically equivalent ways, this latter moment was marked at once by greater anxiety and by the dominance of the language of loss.

No one, it seems, wanted to be responsible for loss. Even Joule—whose researches pointed Thomson so inexorably toward loss—dissociated himself from this position. Indeed, Joule interpreted Carnot rather differently than Thomson did, for he worried that Carnot implied too much loss:

I conceive that this theory, however ingenious, is opposed to the recognized principles of philosophy, because it leads to the conclusion that *vis viva* may be destroyed by an improper disposition of the apparatus. . . . Believing that the power to destroy belongs to the Creator alone, I entirely coincide with Roget and Faraday in the opinion that any theory which, when carried out, demands the annihilation of force, is necessarily erroneous.[81]

That Joule's objection was to the loss of *vis viva* (living force, or kinetic energy), whereas Thomson's was to the loss of mechanical effect or, alternately, to the generation of heat, suggests the definitional challenges central to the consolidation of energy physics, especially the shifting uses of *heat*.

For Joule, it was precisely heat that resolved the threat to conservation. In Joule, as in Tennyson, heat made it possible to reconceive loss as transformation: "Wherever living force is *apparently* destroyed, an equivalent is produced which in process of time may be reconverted into living force. This equivalent is *heat*." Thus reconciled, Joule expressed a faith fully in keeping with the workings of the physical world: "We find a vast variety of phenomena connected with the conversion of living force and heat into one another, which speak in language which cannot be misunderstood of the wisdom and beneficence of the Great Architect of nature."[82] As Joule evoked the metaphor of language, these interconversions became themselves the physical metaphor that attested to the goodness of God. As for Tennyson, "large elements [are] in order brought";[83] loss and disorder prove merely superficial, subordinated to the larger truth: "Thus it is that order is maintained in the universe—nothing is deranged, nothing ever lost, but the entire machinery, complicated as it is, works smoothly and harmoniously . . . everything may appear complicated . . . yet is the most perfect regularity preserved—the whole being governed by the sovereign will of God."[84]

Once Thomson had reconciled Carnot with Joule via the dynamical theory of heat (the theory that understood heat not as substance but as molecular motion), he too reconceived loss—here "waste"—as transformation. And as in Tennyson, this reconceptualization enabled the restoration of faith: "As it is most certain that Creative Power alone can either call into

existence or annihilate mechanical energy, the 'waste' referred
to cannot be annihilation, but must be some transformation of
energy."[85] This assertion marks a moment of transition for
Thomson, regarding the affective resonance of thermodynamics. For this clear statement of faith—now consistent with physical theory—is his opening to a surprisingly short but resonant
piece marking the beginning of the end of his thermodynamic
optimism: "On a Universal Tendency in Nature to the Dissipation of Mechanical Energy."

Diffusing Ambivalence: Sexuality, Gender, and Evolution

This concept of *dissipation*—or its frequent synonym in physical
discourse, *diffusion*—treated Tennyson rather more gently. As
we have seen, the concept of diffusion enabled Tennyson to
conceive of conservation in the face of apparent loss. The poet's
capacity "to feel thee some diffusive power," moreover, suggests
the usefulness of Tennyson's energetics for diffusing the explo-
sive potential of his feelings for Hallam. For all that Tennyson
may "seem to love [him] more and more," the poet's passion
for Hallam, however vast, is not without ambivalence. Identifying the poem's "strategic equivocation," one scholar explores
how *In Memoriam*'s "elegiac mode disciplines the desire it also
enables," at once articulating and sublimating Tennyson's
longing for Hallam.[86] Another locates the poem's homoerotics
as what the Victorians (and Freud) considered a stage in the
sexual evolution of the individual—an early phase that the individual was thought to leave behind on reaching (hetero)sexual
maturity.[87] I would argue that the poet's capacity "to feel thee
some diffusive power"—a capacity that speaks to the conservation of both energy and love—intersects with these important

readings of the text's homoerotics. For this passage is both conservative and dissipative, not only thermodynamically, but also sexually. By diffusing the power of Hallam's attraction, thermodynamics consoles the poet for the loss of his friend, even as it disperses his uneasiness about the nature of his affection. Though not lost, Hallam's power is diffused; it is spread out and therefore less useful, more entropic. Similarly, the poet's passion may be "vaster," but it is certainly less usable. Providing a way for Hallam to survive pointedly disembodied, the mechanics of energy transformation allow the poet to insist that Hallam's attraction is eternal, even as he renders it lukewarm.

Nor is Hallam's the only disconcerting power diffused in this way. Ultimately, Tennyson will also diffuse the power of a fiercely feminized Mother Nature, who is not only "red in tooth and claw," but also ravenous and, though reproductive, distinctly unmotherly. It is no wonder, then, that in his early cosmology, Tennyson retroactively endows "Nature" with power, only in the moment that he imaginatively divests her of that power (recall "Nature's ancient power is lost"). But by the end of the poem, Tennyson manages to diffuse the power of this frightful female without the gloomy sacrifice of spring. When in his diffusion, Hallam becomes "mix'd with God and Nature," he enters an androgynous being already in progress. And this blending is marked by Tennyson's prethermodynamic studies, for, formerly "at strife," when God and Nature do meet, they do so "in light."[88]

Thus, the consolation of physics proves remarkably adaptable, and the principles of simultaneous conservation and dissipation that permeate Tennyson's personal consolation are reiterated in the poem's popular consolation. Conserving Hallam's love while diffusing the power of his attraction, and dissipating

Nature's frightful femininity through a mixing with God, Tennyson's thermodynamic solution—his answer to death and the second law—also heals the rift between science and faith wrought by science's depiction of a ruthless nature. And insofar as this picture of nature reinforces the poem's evolutionary angst, the consolation of thermodynamics enables Tennyson's reformulation of the implications of evolution as well. For if "literature is especially designated in a society . . . to express and shape affective meanings,"[89] *In Memoriam* forces us to revise our picture of an antagonistic relationship between Victorian thermodynamic and evolutionary narrative. A poem driven, at the outset, by a sense of overwhelming loss, *In Memoriam* will be able to produce a narrative of evolutionary progress only through the kind of willed optimism that attaches, as I have argued, to Tennyson's maintenance of a first-law sensibility. Such a narrative must derive from the same capacity to imagine conservation in the face of dissipation, transformation in the place of loss. *In Memoriam* manifests this capacity abundantly—even in its geological narrative. For while early in the poem, geology is marked by waste and decay as the speaker hears erosion in "the sound of streams that swift or slow/Draw down Aeonian hills and sow/The dust of continents to be,"[90] this geological image gives way to another:

> The hills are shadows, and they flow
>> From form to form, and nothing stands:
>> They melt like mist, the solid lands,
> Like clouds, they shape themselves and go.[91]

In this latter image, geology is marked by the dominance of change over loss. Hills no longer erode into dust; they flow from form to form, from a solid to a mist, perhaps, but still present, still shaping themselves, and like clouds, increasingly capable of transformation. The poem's ability to reshape anxieties of (here, geological) loss and dissipation into hopes of transformation and conservation, suggests the power of Tennyson's increasingly present first-law optimism, a discourse sufficiently powerful that it could not fail to inform the poem's evolutionary narrative as well.

As these two scientific subtexts become entangled, physics—counter to late-century expectations—offers a solution to the problems of waste wrought by biology, for in biology, too, the earlier parts of the poem imply no progress and much waste. Manifestly concerned with the careless waste of resources in the reproduction of species, the speaker fixes on the profligacy of Mother Nature and "find[s] that of fifty seeds, she often brings but one to bear."[92] But evolutionary biology, the poem suggests, is able to transform waste into progress because it grows up alongside of and in conversation with the notion of transformation central to the development of thermodynamics—especially its first law. *In Memoriam's* evolutionary thinking follows a pattern that is by now familiar as this early vision of waste is eventually replaced by a vision of progress, a vision wherein man—the species and the individual—"move[s] upward, working out the beast,/And let[s] the ape and tiger die."[93] We can then see in Tennysonian evolution—"evolution as transformation, maintaining identity but bringing about change"—the undertones of the first and second laws of thermodynamics. If "Tennyson's concept demanded transformation,"[94] the nascent principles of thermodynamics provided a mechanism for answering that demand, for finding transformation where the senses perceived loss and waste. Thus, the triumph at the end of the poem of a progressivist evolutionary narrative is identically the triumph of first-law optimism and the fantasy of transformation over second-law anxieties of inevitable loss.

These simultaneous triumphs merge with the renewal of faith and the reconciliation of science and religion on which the poem ends. Transformation explains apparent loss; religious, evolutionary, and thermodynamic optimism converge in the final lines of the poem: "One God, one law, one element,/And one far off divine event,/To which the whole creation moves." This passage marks, once more, Tennyson's debt to and divergence from his Romantic predecessors. Like and unlike Coleridge, Tennyson expresses a belief in divine unity, but one that has become—like Maxwell's "infinite continuity"—markedly scientific. Romantic wholeness has been reconciled with Victorian science. Undoubtedly the confirmation of faith that readers have always taken it to be, the "one far off divine event" is emphatically overdetermined. While it undoubtedly represents a union with God, this union is also figured as the culmination of upward evolutionary progress, a time when, having "moved through life of lower phase," we will evolve into a species "no longer half akin to brute," the "crowning race" for whom Hallam was the herald "appearing ere the times were ripe." But when read in light of the poem's thermodynamic concerns, God comes to look remarkably like a heat sink. No longer "somewhere in the waste" or even lost "in light," the poet decides "that friend of mine [now] lives *in* God."[95] Nature's power safely diffused, God alone remains, the repository of all energy that has passed irrecoverably to the other side.

"One law," moreover, reemphasizes the wish—characteristic of *In Memoriam* and thermodynamics alike—for the widespread application of physical principles. As first-law optimism pervades the poem's cosmology, "one far off divine event" (the flip side of "Spring no more") suggests a moment in which currently interconvertible elements—heat, electricity, work, light, and life—merge as one divine undifferentiated element. A very optimistic take on heat death! Tennyson, then, closes the system, as he closes the poem—both vast enough to encompass what once seemed lost—even as he expresses religious and scientific hope for universal applicability. For the monotheistic mission of "one God, one law, one element" suggests a scientific aspiration as well: the desire, attached to the discovery of energy transformation, that we will discover a single law that governs all natural processes—a grand unified theory for the nineteenth century, a hope that nineteenth-century energy physics will hold close to the heart.

II Energy and Empire: Applications

2 Grand Unified Theories, or Who's Got GUTs?

In our day grand generalizations have been reached. The theory of the origin of species is but one of them. Another, of still wider grasp and more radical significance, is the doctrine of the Conservation of Energy, the ultimate philosophical issues of which are as yet but dimly seen—
—John Tyndall[1]

One Law

In the twentieth century, physics strove to create what was known as a grand unified theory, which would connect what seemed like four distinct basic forces, electromagnetism, strong and weak nuclear forces, and gravitation, such that they revealed themselves as one. Victorian energy physics made a similar effort to unify apparently disparate observables. Eventually heat, light, electricity, work (mechanical effect), magnetism, and motion were all comprehended within *energy*. But for Victorians, apparently, this list was insufficiently comprehensive.[2] Tennyson, as we have seen, brings not only evolution, but also religion into the fold of what these new principles can and should explain. And his "one God, one law, one element" resonates in the claim physicists Thomson and Tait repeatedly make for the conservation of energy—that it represents "the ONE GREAT LAW of Physical Science." Just as their 1862 essay "Energy" insisted on the primacy of the conservation of energy, it also suggested that the science of energy was consistent with religion, quoting the scriptural evocation of a second-law eventuality: "The elements should melt with fervent heat." This is pretty typical, and not at all surprising in scientists steeped (as Thomson was) in the traditions of Scottish Presbyterianism.[3] But such efforts were by no means limited to those whose science was so firmly grounded in faith. Scientific naturalists who hoped to dissociate religion from science were nonetheless tempted by the impulse to unify. After all, if science was to supplant rather than reinforce religion, it would have a lot of explaining to do. If it was to take what has been called positivism's "final step . . . in the pursuit of knowledge," it would at least have to apply "the method of natural science to the elucidation of human behavior."[4] Some thought the principles of energy—already so useful in unifying natural phenomena—should govern anatomy, biology, evolution, psychology, history, politics, economics, society, and the soul. The widespread conversation I've termed *ThermoPoetics* thus attracted not only physicists and poets, but also political philosophers and novelists.

Energy physics had an undeniable appeal for those who sought the one law to rule them all (one law to bind them), who sought the *first principles* from which all others could be derived. This was the pet project of one Victorian philosopher closely associated with scientific naturalism: Herbert Spencer. Best known for his phrase "survival of the fittest,"[5] Spencer engaged in an extended project of energetic unification. Driven by his conviction that "Philosophy is completely unified knowledge," Spencer sought to explain everything from the motion of molecules to the creation of solar systems, the action of individuals and the evolution of species, processes vital, psychological, economic, and governmental. These all derive, he argued in the treatise *First Principles*, from what he called "the persistence of force," something we recognize, more or less, as the conservation of energy. In extending this principle and its consequences to all observable phenomena, Spencer found what he sought. He claimed that "in their ensemble the general truths reached exhibit under certain aspects, a oneness not hitherto observed." And this oneness, according to Spencer, was not just the finding but the imperative of philosophy. The unique task of philosophy was to unify other knowledges, or what should amount to the same thing, to uncover a unity that already exists.[6]

Spencer put forth his strategy as follows: "To unify the truths just reached with other truths," he said, "they must be deduced from the Persistence of Force"; it is "ultimate and the others derivative." Spencer preferred the phrase "persistence of force" to "conservation of energy," but it reflected a very similar principle: "The recognition of a persistent Force, ever changing its manifestations but unchanged in quantity throughout all past time and all future time, is that which we find alone makes possible each concrete interpretation, and at last unifies all concrete interpretations." Spencer's particular spin on conservation thus elevates it to the fundamental principle that explains and binds pretty much everything. The unification of all concrete interpretations effected by philosophy through judicious uses of the persistence of force reveals all aspects of the cosmos to be not just analogous, but at base the same. Within physics this means that heat, light, electricity, work, and so on, are all interconvertible. In Spencer's philosophy, this universal truth should comprehend all phenomena.[7]

One effect of this project is to narrow the already fine line between analogy and identity.[8] Repeatedly, the scientists developing nineteenth-century physical theory evoked analogies between phenomena that would eventually be subsumed within the category of energy. Without in any way holding to the theory of caloric—a theory of heat that takes it to be a literal fluid—James Thomson would evoke the analogy of a waterfall to describe the workings of heat energy. James Clerk Maxwell would draw on the analogy of fluid flow to develop his equations of electricity and magnetism. And William Thomson would construct "a mathematical analogy between electrostatic induction and heat conduction"—all part and parcel of his larger quest for a unified field theory, rendered urgent once Thomson adopts his dynamical theory of heat.[9]

With *First Principles*, Spencer engaged in a similar process of analogy making. Now, Spencer was not a physicist. He didn't sweat the physical evidence or the mathematical rigor. But as he described the origin of solar systems, the growth of bodies, the evolution of species, and the development of societies, these looked increasingly similar—analogous, one might say. And Spencer claimed for them all a common cause: the persistence of force, of course. Relying thus heavily on the use of analogy, Spencer's philosophy evoked not only the work of scientists, but also the work of fiction. This is not to say that he was just making it up—though undoubtedly there are elements of the fictional in his work—but that the project in which he was engaged is one that lent itself to fictional treatment. To see how this works, let's have a look at Spencer alongside a fictional work dedicated to a similar project: the novel *The Coming Race* (1871) by Edward Bulwer, Lord Lytton.[10]

Do not fret if you've never heard of *The Coming Race* or even Bulwer-Lytton. He was a Victorian novelist who gave us such expressions as the "almighty dollar" and "It was a dark and stormy night." In fact, the latter has earned the author lasting fame through the Bulwer-Lytton Fiction Contest in which writers compete "to compose the opening sentence to the worst of all possible novels."[11] But *The Coming Race* is a very interesting example of early science fiction, involving the discovery of an advanced race of people, the Vril-ya, who have evolved underground and who possess a remarkable facility with something they call *Vril*, a concept akin to what we call *energy*. What's more, *The Coming Race* was "an immediate best-seller" and has been called the "crowning achievement" of the hollow-earth subgenre, and one, moreover, that "is unequaled for the depths of its intellectual explorations—inquiries into an astonishing range of social, political, scientific, religious, linguistic, and sexual issues."[12] And so, with Spencer and Bulwer-Lytton, we extend our thinking about poetry and physics to philosophy and fiction. These are all, I contend, engaged in conversation. *First Principles* and *The Coming Race* prove remarkably like-minded; their authors were clearly thinking about similar stuff. I am not, by the way, suggesting that one was simply responding to the other, that Bulwer-Lytton simply put Spencer's ideas into fictional form, or anything like that. As far as we know, they didn't collaborate and they weren't particularly friends. They had contact,[13] but I would have written this chapter even if they hadn't.

They certainly agreed, however, that the fundamental function of philosophy is to unify. Bulwer-Lytton explained this principle (more simply than Spencer seems able to manage) as he described the advanced people he imagined in his novel, the "tribe of Vril-ya." Their government was "apparently very complicated, really very simple. It was based upon a principle recognized in theory, though little carried out in practice, above ground—viz., that the object of all systems of philosophical thought tends to the attainment of unity." Somehow, the Vril-ya manage, as we cannot, to "[ascend] through all intervening labyrinths to the simplicity of a single first cause or principle."[14] Inasmuch as Bulwer-Lytton described an apparently utopian government, the "attainment of unity" suggests the unity of people(s) under a benign and minimalist governing body. But, like Spencer's, this unity also refers to how we simplify our thinking by identifying a single first principle. Unlike Spencer, Bulwer-Lytton was not specific about what that first principle was, though subsequent comparisons to machinery suggest a physical principle. Meanwhile, Bulwer-Lytton's simultaneous assertion of philosophical and political unity as well as the comprehensive phrasing "all systems of philosophical thought" suggest that a larger intellectual unity is at stake, something that may comprehend the physical, the governmental, and the philosophical—a truly grand, unified theory. *The Coming Race* further suggests its like-mindedness to *First Principles* in its efforts to unify its own systems of philosophical thought. As it lays out the various aspects of the culture of the Vril-ya, a sameness emerges that suggests an attempt to fictionally test the viability of energy physics as the source of its own first principles.

Physics and Evolution

Like so many Victorian texts, *The Coming Race* and *First Principles* seem obviously concerned with evolution.[15] Spencer's broader associations with evolution and, of course, social Darwinism, as well as his diehard progressivism, his preferred vocabulary, and his association with scientific naturalism, keep such readings foremost. In Bulwer-Lytton's case, it is probably because his narrative takes place underground among an advanced species of humans who claim to be descended from certain remarkably intelligent and very large frogs. But with Bulwer-Lytton, as with Spencer, evolution itself proves to be a specific case of energy principles at work.[16]

Spencer works hard to develop what he calls "a complete unification . . . a synthesis in which Evolution in general and in detail becomes known as an implication of the law that transcends proof." This law is the persistence of force. And Spencer's assertion that evolution derives from it, does not clearly explain *how*, except to say that all processes of change, including evolution however construed, "are so many different aspects of one transformation."[17] It takes most of *First Principles* to establish, as it were, the identity of all transformations of matter and motion, from the action of molecules to the evolution of life, the course of civilization, the origin and demise of galaxies. The behavior of matter in motion under the persistence of force is thus to account for all transformations. According to Spencer, it is all about the way matter clumps, becoming less evenly spread out, more localized and complex. Spencer further insists that matter becomes more "coherent"—a word that implies not just that it sticks together, but that it somehow makes sense, that it is more orderly. As Spencer comes to define evolution in general, his terms are those of physics. He describes evolution as "an integration of matter and concomitant dissipation of motion." Matter, he says, "passes from an indefinite, incoherent homogeneity to a definite, coherent heterogeneity." He will actually have some trouble making the case for anything so counterentropic. Ultimately, he will concede that dissolution is as likely as evolution. But even as he defines evolution, he adds the caveat that anticipates these later gestures, that "the definition of Evolution needs qualifying by introduction of the word 'relatively' before each of its antithetical clauses."[18] We cannot talk about absolute beginnings and endings—any more than Darwin's "origin" of species can account for the very inceptions of life. Both refer only to processes in medias res, systems developing from other systems. We will return to Spencer's mechanism for coping with the law of entropy. For now, it is important to note that Spencer's evolution, both in general and in detail, is about producing order from disorder in a manner consonant with the laws of physics. It is a secondary process, a consequence of the persistence of force. And though very different from *In Memoriam*, *First Principles* recalls Tennyson's version of evolution as transformation in this: building on a conservation principle, Spencer moves from an image of waste—an indefinite, incoherent homogeneity—into the building blocks of progress, the development of a definite, coherent heterogeneity.

As in Spencer, there is plenty in Bulwer-Lytton's fantasy of alternative evolution to satisfy our sci-fi predilection for ficto-plausible explanation. There may even seem little at first to make us suspect that more is at stake than Darwinian biology. Bulwer-Lytton himself notes, in a letter to his editor, that "the only important point is to keep in view the Darwinian proposition that a coming race is destined to supplant our races."[19] As the novel accounts for the evolution of the Vril-ya, we get nothing so much as a Spencerian-style discourse on the "struggle for survival." The brilliant and forbidding Zee, a woman and sage of the Vril-ya, explains that when "life is made a struggle, in which the individual has to put forth all his powers to compete with his fellow," we "invariably find" that "in the competition a vast number must perish, nature selects for preservation only the strongest specimens." Like Spencer, Zee draws social implications from Darwinian principles, locating them in "the early process in the history of civilization."

But as in Tennyson and Spencer, we find evolution linked to, even driven, by energy physics. We hear echoes of the profligacy of nature and the transformation of waste into progress so characteristic of Tennysonian evolution. "Even before the discovery of vril," Zee assures us, "only the highest organizations were preserved." Zee's phrase "highest organizations" suggests increasing order and complexity, like Spencer's "definite, coherent heterogeneity." And environmental energy concerns are implicit in the conditions that intensify the early struggles of the Vril-ya, accelerating their evolution relative to that of humans—for the Vril-ya have no easy-to-use, localized energy source. They live where the sun doesn't shine. And so enters vril. Its name evokes a (gendered) notion of species-wide virility. It is a discovery that propels the Vril-ya far ahead of all other species and serves to distinguish them from the nations they consider barbarians. They even develop a specialized nerve in the hand, without which our hopelessly barbaric human narrator can never hope to "achieve other than imperfect and feeble power over the agency of vril." Vril is as near to energy as we could wish. And the Vril-ya insist that their facility with it makes all the difference.[20]

Energy Unifies

It is this concern with energetics that appears first in the novel and provides the context within which we must understand its evolutionary narrative. Our narrator is a hapless American who literally drops in on the Vril-ya by way of a mine. One glance at his future host tells him that "this manlike image was endowed with forces inimical to man."[21] And the explanation provided by his hostess soon reveals what these "forces" are:

Therewith Zee began to enter into an explanation of which I understood very little, for there is no word in any language I know which is an exact synonym for vril. I should call it electricity, except that it comprehends in its manifold branches other forces of nature, to which, in our scientific nomenclature, differing names are assigned, such as magnetism, galvanism, &c. These people consider that in vril they have arrived at the unity in natural energetic agencies, which has been conjectured by many philosophers above ground, and which Faraday thus intimates under the more cautious term of correlation:

"I have long held an opinion," says that illustrious experimentalist, "almost amounting to a conviction, in common, I believe, with many other lovers of natural knowledge, that the various forms under which the forces of matter are made manifest, have one common origin; or, in other words, are so directly related and mutually dependent that they are convertible, as it were into one another, and possess equivalents of power in their action."[22]

This introduction foregrounds the unification principle that attaches to the development of energy physics. Many different phenomena—electricity, magnetism, galvanism, and potentially a host of others—are all comprehended within *vril* as they will soon come to be under the term *energy*. With *vril*, the Vril-ya have arrived at a "unity in natural energetic agencies" still in process above ground and loosely marked by Faraday's term *correlation*, the concept of "common origin," and his belief in the convertibility of forces and "equivalents of power." Even for Faraday, these physical principles also effect a social unification, uniting him through common conviction "with many other lovers of natural knowledge."[23]

Similarly, as the Vril-ya effect a conceptual unification, energy returns the favor, uniting the nations of the Vril-ya in one supercommunity that dominates the world underground. This happens quite naturally, the government itself operating "so noiselessly, and quietly that the evidence of a government seems to vanish altogether, and social order to be as regular and unobtrusive as if it were a law of nature." Nature thus becomes fully intertwined with both culture and technology. As Bulwer-Lytton describes the government of the Vril-ya, it becomes clear that it runs like—as well as on the same principles as—a well-oiled machine; the government's employment of machinery merely extends the principles at work throughout the social order. Within the Vril-ya government, "machinery is employed to an inconceivable extent in all the operations of labour within and without doors, and it is the unceasing object of the department charged with its administration to extend its efficiency."[24] This effort to extend the efficiency of machinery or of government recalls the roots of energy principles in the drive for economic efficiency. James Thomson, Sadi Carnot, James Watt, and others all hoped to build better engines because these made better use of resources, did more work, cost less, or earned more for the individual or nation that could boast them. In Bulwer-Lytton's hands, the analogy between the workings of government and the workings of machinery becomes more nearly an identity. Both are governed as by the law of nature; no other laws are necessary or possible.

Perhaps because he too comes from a background in engineering, Herbert Spencer is similarly tempted to apply physical principles to a wide range of—ideally all—systems, extending the persistence of force to large and small physical systems, to biology and geology, psychology, sociology, and history. As he puts it, "If the general law of transformation and equivalence holds of the forces we class as vital and mental, it must hold also of those which we class as social." This principle, of course, pertains to the "physical forces that [may be] directly transformed into social ones." All of these transformations "inevitably follow" from the persistence of force, the "proposition that force can neither come into existence nor cease to exist."[25]

With the "gradual discovery of the latent powers stored in the all-permeating fluid which they denominate Vril," the Vril-ya eventually transform physical forces into social ones. At once destructive "like a flash of lightning," it can also "invigorate life, heal, and preserve." And in a sci-fi move that anticipates the Cold War by close to a century, Vril has also brought about peace through the capacity for total annihilation. The "unity in natural energetic agencies" that is Vril thus effects transformations that we class as vital and social. (Elsewhere, we see mental transformations as well.) Perfecting the destructive capacity of Vril, its discoverers eliminate "all notions of government by force." However precarious, this particular means of securing peace is consistent with the central principle of Vril-ya physiology: health (in all things) depends on establishing "due equilibrium of . . . natural powers." Both Spencer and the Vril-ya see equilibrium as the necessary and proper end of energy principles applied to their fullest extent. And as we will see, both the Vril-ya and Spencer eventually allow the principle of equilibrium to dominate their energetics and, what amounts to the same thing, their worldviews.[26]

The Fantasy of Equilibrium

Among the Vril-ya, this principle of equilibrium manifests itself in social structure, in religion, and even in the way they light their streets. The Vril-ya strive for an equilibrated culture, one that eschews distinctions of rank and moral hierarchy and that eventually substitutes "serene equality" for all government by force. They seek repose or rest. Spencer, on the other hand, grounds his theories of evolution, especially social but biological evolution as well, in his "doctrine of ultimate equilibration." Fully compatible with the "persistence of force," it posits a time when all forces are in balance or in a state of perfect exchange. For both, as we will see, the principle of equilibrium proves a difficult doctrine to sustain.

As Spencer pursues the "persistence of force" to every logical consequence he can, applying the principle to the development of systems solar, social, physiological, biological, psychological, governmental, and so on, he finally raises the question of where this all tends: "At length, to the query whether these processes have any limit, there came the answer that they must end in equilibrium." He further specifies the application of this end to our world and species: "And our concluding inference was, that the penultimate stage of equilibration in the organic world, in which the extremest multiformity and most complex moving equilibrium are established, must be one implying the highest state of humanity."[27] I'll return to the question of why this is the "penultimate stage" of "moving equilibrium." Meanwhile, it is enough to note that all processes that result from the persistence of force—which is to say, all processes—must end in equilibrium. For Spencer, entropy takes the form of "that gradual dissipation of motions . . . and the consequent tendency towards a balanced distribution of forces. . . . In their totality, these complex motions constitute a dependent moving equilibrium."[28] This is Spencer's spin on what will solidify as the first and second laws of thermodynamics. The gradual dissipation of motion represents not so much the loss as the redistribution of forces. As what others call entropy increases, these forces come closer to an even or balanced distribution.

Intermediate stages, all of which may be classed as forms of evolution, are in the process of equilibration. Socially, we may see this at work not only in something we call progress, but also in the very fluctuations that are adjustments toward equilibrium: "Each society displays the process of equilibration in the continuous adjustment of its population to its means of subsistence." During times of plenty, people have more children—or as Spencer delicately puts it, "the proportion of marriages increases." It goes down, of course, when provisions are scarce. For the reproductive capacity of a community, Spencer actually uses the phrase "expansive force"; "repressive force" denotes things such as space and resources that check increases in population. And when he ties them together, he sounds positively Newtonian: "It will be seen that the expansive force produces unusual advances whenever the repressive force diminishes, as *vice versa*." This continued use of *force* when discussing reproduction evokes Spencer's pet principle—the persistence of force—even as it broadens the potential implications of expansive and repressive forces. Human reproduction gives way seamlessly to artificial production, and we find that the limit on industrial progress is another form of equilibrium, that which "Mr. [John Stuart] Mill has called 'the stationary state.'" Once we have filled up the globe with people, explored every resource, and perfected the "productive arts," the outcome must be "an almost complete balance, both between the fertility and mortality in each society, and between its producing and consuming activities." In Spencer's words, this stationary state indicates a mature culture, whose "various industrial activities settle down into a comparatively constant state." The more advanced the culture, the more it is conducive to such equilibration. Good transportation and communication technologies enable the even distribution of people, information, and goods. Supply meets demand. Prices fluctuate less. People are born and die at roughly the same rate. Government is affected by the same principle of balance: "Governmental institutions . . . fall into harmony with the desires of the people." It is thus, according to Spencer, that a culture approaches equilibrium.[29]

The Vril-ya provide a lovely example of such a mature, equilibrated culture, characterized by imperceptibly slow progress and careful balance. It is evident not only in the balance of population with resources—fertility and mortality, producing and consuming—but also in the particular ways they use those resources, conserving (in the colloquial, if not the technical sense) both physical and social energy, and keeping a balanced distribution of population, rank, effort, and even light. Without of course putting force on individual inclinations, the Vril-ya attempt to balance the resources of each region with the needs and wants of the population. "In the course of a few generations, [the Vril-discoverers] peacefully split into communities of moderate size," and to maintain a size appropriate to its territory, "at stated periods, [a volunteer corps] of surplus population departed to seek a realm of its own."[30]

In this way, the Vril-ya seem to have reached a form of equilibrium. Though they have not reached that final state, where population is dense everywhere and the resources of every underground region have been fully explored, they do see the value of fitting producing to consuming activities. Within each community, the Vril-ya seem to enjoy something very like Mill's stationary state. "After centuries of struggle" they have, in words that recall Spencer's, "settled into a form of government with which [they] are content." They "allow no differences" and use "no insignia of rank."[31] They pay "no honours . . . to administrators distinguishing them from others."[32] And, because there is "no difference of rank or position between the grades of wealth or the choice of occupations, each pursues his own inclinations without creating envy or vying."[33] In this way, the Vril-ya approximate that stationary state in which the productive arts admit of no further improvement, by limiting incentives to such arts: "There is no stimulus given to individual ambition. No one would read works advocating theories that involved any political or social change, and therefore no one writes them."[34] The close association made here between difference(s) or distinction(s) and (stimulus to) change suggests the physical principle that drives all heat engines and more broadly, enables work. I will return to this principle with *Bleak House*. Meanwhile, it suffices to note that as those differences are gradually removed—as heat goes to uniform temperature or water seeks its own level, or, alternately, as rank is dissolved—work or change is no longer possible. For the Vril-ya, it is also no longer desirable. A Vril-ya proverb further connects such leveling to unity, order, and happiness: "No happiness without order, no order without authority, no authority without unity."[35] The progress toward unity, at the expense of individual distinction, appears in Spencer's account of societies advancing as well. "For," he says, "among associated men the progress is ever towards a merging of individual actions in the actions of corporate bodies."[36]

Diffusion

This state of affairs comes about in part through a process of rapid *diffusion*—a word that signals at once the intermingling of particles due to thermal motion, the scattering of light, the spread of social institutions, and (I'll add ominously) *dissipation*. In all cases, diffusion is part of the process of equilibration. Spencer observes that when, in primitive cultures, "the diffusion of mercantile information is slow," supply cannot accommodate demand; wide market fluctuations and other such non-equilibrium fluctuations result. When there is "rapid diffusion of printed or written intelligence," railways, telegraphs, good roads, and all other things come into balance.[37] People, goods, and information flow freely, spread out evenly, equilibrate. Similarly, the Vril-ya value diffusion for its equilibrating effects. It is the means by which civilization is sustained and streets are lit. Indeed, our host defines *civilization* as the "art of diffusing throughout a community the tranquil happiness that belongs to a virtuous and well-ordered household." In both Bulwer-Lytton and Spencer, we are meant to understand this diffusion as conducive to order. It also signals a more advanced state of affairs. Among the Vril-ya, this diffuse social order is reflected in and reinforced by the light they maintain for their underground world. The narrator notes the peculiar quality of their light as soon as he arrives: "It seemed to me a diffused atmospheric light, not like that from fire, but soft and silvery, as from a northern star." It is at once pleasant and even and seems to manifest the advanced state whereby the Vril-ya have substituted equilibrium for force. As does the heat: "The world without a sun was bright and warm as an Italian landscape at noon, but the air less oppressive, the heat softer."[38] Both the light and heat of the Vril-ya are softer, kinder, gentler than those above ground. Diffuse, even cold, as from a faraway star, rather than localized, hot and intense as from a nearby source, a fire, or even the sun, this light suggests something not only softer, but also more advanced, older, and in spite of its associations with order, more entropic.

Living Force

The Vril-ya thus maintain "to the limits of their territory, the same degree of light at all hours." The lighting of their communities serves not merely as metaphor (or even "pathetic fallacy") to represent their desire for equilibrium, but as a basic component of the widespread application of energy principles. Through the power of Vril, they not only light their streets, but also control the weather "by one operation of vril, which Faraday would perhaps call 'atmospheric magnetism.'" The influence that Vril can exercise over the mind—through "operations, akin to those ascribed to mesmerism, electro-biology, odic force, &c., but applied scientifically, through vril conductors"[39]—is demonstrated repeatedly on our narrator, who is variously sent to sleep, placed in trances, taught a new language, and physically driven through the agency of Vril. But lest we dissociate these effects from their physical causes, we are reminded of the fundamental physics at the base of Vril-ya power, philosophy, and technology as Zee expresses her incredulity at the ignorance of the narrator. "Your parents or tutors surely cannot have left you so ignorant," she says, "as not to know that no form of matter is motionless and inert." In words that would make John Tyndall (author of the popular book *Heat: A Mode of Motion* [1863]) proud, she continues, "Every particle is constantly in motion and constantly acted upon by agencies, of which heat is the most apparent and rapid." Of course, the Vril-ya are not limited by mere human science. They know further that of such agencies, "vril [is] the most subtle, and, when skilfully wielded, the most powerful."[40] This understanding of the internal (aka molecular) motion of matter leads not only to highly advanced technology, but also to a unification of religion and science, life and nonlife, body and mind.

The Vril-ya enjoy automation at a level of sophistication still possible only in science fiction. But science fiction has the capacity to give life to metaphor, to *literalize*.[41] *The Coming Race* animates the metaphor always present in the physical term *vis viva*, literally "living force," or what we now call *kinetic energy*, the energy of motion. As it happens, the particles (atoms, molecules) that make up matter are in constant, random motion. Each of these has a quantity of kinetic energy. The sum of individual kinetic energies is called *thermal energy*. We call the transfer of thermal energy *heat*. It is through heat (or thermal energy transfer) that we sense the internal movement to which Zee attests, "the action which is eternally at work upon every particle of matter." Of course, the Vril-ya can do more with it than we can. Through "sufficient force of the vril power," the Vril-ya can control and increase the internal motion of matter, such that any "heap of metal" can be "as much compelled to obey as if it were displaced by a visible bodily force."[42]

To the Vril-ya, the random motion of molecules constitutes the life or potential for life of even matter at its most "inert" (which suggests nonlife) and "stubborn" (which by humanizing matter already suggests life). Now if the potential for life inheres in vis viva, the energy of motion, we shouldn't be surprised if it is subject to a law of conservation. And among the Vril-ya—as among us—conservation figures first as a religious principle. In terms that should remind us of Joule's conviction that we can never create or destroy the "powers with which God has endowed matter," the Vril-ya "hold that wherever He has once given life

.

[it] is never destroyed; it passes into new and improved forms."
Before living force has been reframed as kinetic energy, Joule is
convinced that "such absolute destruction of living force cannot
possibly take place."[43] As life again announces its similarity to
energy, by showing itself subject to a law of conservation, "He"
gives life much as the "intellectual agent" animates the automaton. In both cases, we may doubt that we are seeing life. In the
case of a heap of metal, "one may almost say that it lives and
reasons." In the case of a plant (Zee explains) our perceptions
of life may be faint, but the life is nonetheless present and sub-
jected to the same law of conservation as life that is easier to
see. That the life in the plant is merely faintly perceptible suggests a continuum between what is called life and what is *almost*
called life. It suggests that the distinctions are matters of perception and articulation; the lines are hard to draw. What is
clear is that energy principles govern and unite both life and
nonlife.

And then there is soul. The effort to unify extends even to
gestures within *The Coming Race* to make sense of the spiritual
through the principles of energy. Zee's explanation that the
automaton "is animated for the time being by the soul thus
infused into it"[44] suggests that "soul" operates as the natural
force produced by harnessing the internal motion of matter.[45]
It is not clear, and perhaps not important, whether this is an
attempt to naturalize soul or to spiritualize energy, but it certainly feels familiar. P. G. Tait (whose name we have already
seen coupled with Thomson's) and Balfour Stewart (another
scientist working on things like magnetism and heat) published
an extended attempt to marry the physical to the spiritual via
energy in *The Unseen Universe or Physical Speculations on a Future
State*. Purportedly driven only by scientific principles, they argue

for the presence of an Unseen Universe, "full of life and intelligence . . . in fact a spiritual universe and not a dead one." Beginning with the Egyptian belief that "nothing perishes which has once existed" and working through a history of similar articulations of conservation, they posit the Unseen Universe as the repository of energy beyond our reach. Ultimately, they are "led to a scientific conception of [the universe] which is strikingly analogous to the system which is presented to us in the Christian religion."[46] And so we move back to Herbert Spencer, because even though Stewart and Tait exhibit a religious agenda quite antithetical to Spencer's scientific naturalism (what Stewart and Tait would call the "extreme scientific school"[47]), *The Unseen Universe* converges with *First Principles* on the subject of unified theories. Stewart and Tait end roughly where Spencer begins, with the conviction that "science and religion neither are nor can be two fields of knowledge with no possible communication between them."[48]

First Principles, in its turn, provided scientific materialists like John Tyndall and Thomas Huxley "a positivist alternative to the natural-theological schemas of Maxwell, Thomson, and Tait."[49] And in spite of the strange-bedfellows feel, Spencer's effort to connect science and religion is at one with his aim to unify theories. Thus, like the Vril-ya, he is led to link the motion of matter to some greater Power, asserting that "Matter, Motion, and Force are but symbols of the Unknown Reality." Outside the natural world that we observe, Spencer posits "a Power of which the nature remains for ever inconceivable [that] works in us certain effects." These effects are the things we observe. They "have certain likenesses of kind, the most general of which we class together under the names of Matter, Motion, and Force."[50] If Spencer differs from Stewart and Tait on what we can conceive,[51] Spencer's Unknown Reality nonetheless shares many characteristics with their Unseen Universe, including visible manifestations in the same physics of matter, motion, and force. That "we class together" the effects of Spencer's Power points to the broad impulse to unify physical and even spiritual phenomena, to develop from our shared experience "the deepest truths we can reach": our unified theories.

It's a fortuitous impulse, since the business of observing, naming, and classing these effects enables us to distinguish ourselves from Bulwer-Lytton's automata, who seem to live and reason. Otherwise, certain disturbing similarities emerge. That Spencer's Power "works in us certain effects" renders us very like the automata of the Vril-ya—merely matter set in motion by some force beyond our comprehension, not unlike the "intellectual agent" who animates the heap of metal. We have, moreover, already witnessed how our narrator is subject to the will of his Vril-wielding hosts. And if we find this association a bit disturbing, it is not only because we dislike thinking of ourselves as automata, but also because Bulwer-Lytton raises the specter of compulsion by force. For the "soul thus infused into" such a creature compels its obedience "as if it were displaced by a visible bodily force. . . . Without this [the Vril-ya] could not make [their] automata supply the place of servants."[52] Bulwer-Lytton thus anticipates the less benign conversions made possible in the context of energy conservation. Here we witness the conversion of thought into bodily force. One does the work of the other, compelling the automata or the narrator to obey. That *thought* among the Vril-ya is elsewhere converted into firepower emphasizes the double entendre inherent in "force"— and the less humans seem capable of thought and reason, the more easily they're guided by the hand and will of the Vril-ya and thus the greater the possibility that we may "supply the place of servants" on a broader scale as well.

Force Divides

The Vril-ya, of course, disclaim such uses of force, personal and political, insisting that "all notions of government by force gradually vanished from political systems and forms of law." Force, they imply, is necessary only for the maintenance of empire, for "it is only by force that vast communities, dispersed through great distances of space, can be kept together." Because the Vril-ya limit the size of their communities, and have, they claim, nothing "to make one state desire to preponderate in population over another," force becomes unnecessary. They further assert that "no force is put upon individual inclination"; only machines are subject to such compulsion. But the Vril-yan extension of physical laws to other "systems of philosophical thought" makes such distinctions hard to maintain. What constitutes a machine where automata have souls and governments operate according to the laws of nature? And force persists in the "mightiest agency over all forms of matter, animate or inanimate" in the "terrible force of Vril" itself, which can destroy instantly and at a great distance.[53] And so the Vril-ya achieve peace. In fact, nothing has served their peace so well as their capacity for total self-annihilation—unless, perhaps, it is their plans for our annihilation.[54]

Spencer, too, reminds us that force is not easily eliminated from physics or, as it happens, government. And the persistence of the term *force* cannot help but color the attempt to unify theories. Even as energy accompanies a drive to unification, force retains traces of a continued and enforced division, as much between lovers of natural knowledge as between nations. Spencer retains the use of the term *force* in carefully articulated resistance to *energy* and its implications—both intellectual and academic. *Force* not only signals his association with scientific naturalism, but also provides him with a term more in keeping with other philosophical commitments. The term, he believes, implies something broader than *energy*, indeed something that unites energy with, and distinguishes energy from, something else. He defines *force* as comprehending both the "continuity of Motion" and the "indestructibility of Matter." Put differently, Spencer distinguishes two kinds of force, both of which persist: "the force by which matter demonstrates itself to us as existing, and the force by which it demonstrates itself to us as acting." This last is "of the kind known as Energy—a word applied to the force . . . possessed by matter in action." But *energy*, according to Spencer, completely misses that kind of force "by which matter maintains its shape and occupies space: a force which physicists," careless as they are, "appear to think needs no name." As Spencer continues, his "persistence of force" comes to affirm a deeper commitment to what less attentive others call the *conservation of energy*. The term *conservation*, he insists, implies "a conserver and an act of conserving." And if that weren't bad enough, it does not adequately imply that the force existed before a "particular manifestation" and, worse still, seems to assume that "without some act of conservation, force would disappear." *Persistence*, on the other hand, suggests that the force or energy in question existed before, will always exist in some form, and requires no agency to see to it that it persists.[55] As *force* distinguishes itself from *energy*, Spencer thus distinguishes himself from more careless users of language and the North British School of energy physicists. And as careful as he is to thus define force, we cannot casually pass over his reference to "an anti-social force tending ever to cause conflict and separation."[56] These forces, too, are not mere analogues to physical forces. Spencer's unified theory identifies them all as deriving from the "persistence of force"—a phrase with social as well as physical implications. Like other forces, they show strong tendencies to divide.

Force persists as the preferred term among a number of others as well, though by the time of the publication of *The Coming Race*, what Crosbie Smith calls the North British School—including Kelvin, Joule, Tait, and Maxwell—has embraced the conservation of *energy* as the grand unifying principle. Even Mill decries force as a logical fiction.[57] However, force—or more specifically, Tyndall's "On Force"—serves as "the spark that lit the fuse of conflict between rival natural philosophers." If the spark comes from Tyndall's overlooking the contribution of the North British School, especially William Thomson's, in favor of Julius Robert Mayer, a German physician, then the fuel comes from his use of *force* itself: "Tyndall's 'Force' . . . threw down a fresh and potentially very serious challenge from the metropolis. As Huxley's staunchest ally he was preparing to deploy 'conservation of energy' in the cause of scientific naturalism."[58] *Force*, it seems, is a fightin' word.

Almost universally admired, Faraday keeps fairly clear of this conflict. Using *force* freely—in his "lines of force" and "correlation of forces"—he resists Maxwell's wish to make verbal distinctions that not only clarify, but also divide, and eventually ensure the ascendancy of *energy*. A copy of his "On the Conservation of Energy" prompts "a lengthy and positive reply [in which] Maxwell urged that we 'keep our words for distinct things more distinct.'" Maxwell would prefer to reserve *energy* for the "power a thing has of doing work" while defining *force* as the "tendency of a body to pass from one place to another." Faraday replies that he didn't mean force that way, but rather as the "*source* or *sources* of all possible actions of the particles or materials of the universe [or what are] often called the *powers* of nature."[59] But in spite of these linguistic differences, Faraday's work nonetheless represents an important step to move away from action-at-a-distance theories, a step that enables so much subsequent work in the science of energy. It is true that in contemporary usage, energy is a thing conserved; force is not. But Faraday's "correlation of forces" and "conservation of force" are undeniable precursors to the conservation of energy.

Does Bulwer-Lytton's evocation of Faraday and the correlation of forces suggest, then, the tactful choice of a figure everyone can get behind or a resistance to the new terminology? Certainly, he uses the term *energy* far less than Spencer, who objects to it rather strenuously. For all of *The Coming Race*'s concern with Vril, *energy* is used only once, and then only to note that the magistrate's job requires "no preponderant degree" of it. But self-conscious or not, the choice of the cautious, well-known, and widely respected Faraday as an exemplary human enables Bulwer-Lytton to draw a clear distinction between the state of our knowledge and that of the Vril-ya. For in spite of the widespread respect paid to Faraday, his inability to arrive "at a unity in natural energetic agencies,"[60] as the Vril-ya have done, marks the primitive nature of his own culture in the context of *The Coming Race*. Still juggling electricity, magnetism, galvanism, cautiously tiptoeing about the "forces of matter" and making claims for what could be "correlation," "common origin," or "equivalents of power," Faraday, as the first among us, reveals the human failure to fully unite these phenomena. We are still at the stage of conjecture. This less advanced state is further marked by the awkward nomenclature that shows we don't really get it. That our narrator knows of "no word in any language" that does the job of *vril* points to the ongoing struggle taking place above ground (and yes, in reality) to settle the terminology.

Our narrator's unlikely ignorance of the new uses of *energy* portrays above ground physics as being in a rather sorry state. And Bulwer-Lytton takes Zee's description of Vril-ya history as an occasion to critique not only the current state of energy physics, but also the behavior of above ground academics. This, she finds obsolete to the point of being primitive. "In the Wrangling Period of History," she tells us, "whatever one sage asserted another sage was sure to contradict. In fact, it was a maxim in that age, that the human reason could only be sustained aloft by being tossed to and fro in the perpetual motion of contradiction."[61] Wrangling, it seems, is as passé as believing in perpetual motion; any truly enlightened culture knows that neither will sustain anything aloft. And if this condemnation seems merely to suggest that wrangling is *so eighteenth-century*, later descriptions of above ground culture are rather more harsh. The perpetual motion of contradiction finds its widespread counterpart in the "perpetual contest and perpetual change" typical of all those the Vril-ya identify as "savages." We immediately recognize a common Victorian attitude in Zee's condemnation of the savage Other. But then Zee condemns institutions as ignorant and barbaric that are not just familiar but even admired above ground, and we recognize ourselves in her savages, who live "chiefly in that low stage of being, Koom-Posh, tending necessarily to its own hideous dissolution in Glek-Nas." Knowing Koom-Posh to be the word for Democracy, the American narrator is unsettled by Zee's conviction of its inevitable tendency toward dissolution in Glek-Nas, which he translates as the "universal strife-rot." This is, it seems, the closest Vril-ya term to *entropy*, but its implications are generally political. It apparently describes the stage of history that immediately follows democracy gone mad and marks the beginning of the end, "as (to cite illustrations from the upper world) during the French Reign of Terror."[62] It is what happens when the pretense of democracy reveals a glaring disparity. That this disparity—otherwise "contest," "division," "difference," or "distinctions"—is associated with change is also significant, for change, the law of entropy tells us, tends necessarily toward dissolution.

Meanwhile, even for Spencer, the tendency toward conflict reveals a lingering primitivism that must be understood in the terms of physics: "Those aggressive impulses inherited from the pre-social state . . . constitute an anti-social force tending ever to cause conflict and separation." Spencer's evocation of an "anti-social force" is particularly striking coming from someone who is so emphatic regarding the precision and importance of the term *force*. And *force* in a social context works just like a physical force, causing "wide oscillations" in primitive nations between "rigid restraint" or "unbearable repression" on the one hand, and "rebellion and disintegration" or "a bursting of bonds" on the other. In more advanced—that is, better equilibrated—nations, we assume the oscillations are less wide, but we nonetheless "always find violent actions and reactions of the same essential nature."[63] The essentially physical processes that drive political affairs are the same in advanced nations as in primitive ones: action, reaction, force. Spencer's terminology ties his formulation expressly to Newton—a move critical to anyone hoping to achieve legitimacy in Victorian physical science.[64]

Moreover, where Spencer's "despotism" and "unbearable repression" imply something like Bulwer-Lytton's "glaring disparity," we once again see difference driving change and disintegration. Now, the reader may blanch, but the narrator blithely persists in interpreting Koom-Posh, along with rapid change, as progress. As he describes the upper world, he touches "slightly, though indulgently, on the antiquated and decaying institutions of Europe, in order to expatiate on the . . . glorious American Republic." He takes, moreover, as his example, "that city in which progress advances at the fastest rate," as he indulges "in an animated description of the moral habits of New York."[65] To the energetically savvy reader, the narrator's words already reveal the irony of any idea of progress under the second law. In the context of Glek-Nas or social entropy, there can be no difference between the tendency of states he calls "antiquated" and "decaying" and those whose "grandeur" and "preeminence" is measured by rapid progress. Europe "tremblingly foresees its doom" in the United States, but the reverse is also true. Indeed, the United States moves at a faster pace toward its own doom. And if the reader has retained any faith in progress, that faith is undermined by the reference to the "moral habits of New York"—a phrase that does not come off as the compliment the narrator intends. Forward movement must lead to dissolution, and we can only interpret these habits as dissipated, dissolute, or at least dissolving. Finally, even our narrator reveals something of a thermodynamic sensibility, when he observes that the old states, whose "inevitable tendency . . . is towards Koom-Posh-erie," are "gradually shaping their institutions so as to melt into ours."[66] He suggests, without meaning to, that the decay of a culture is a process that implies melting, the loss of differentiation in the increase of entropy.

Thus *The Coming Race* transforms "progress" into decay. The move, moreover, evokes William Thomson's own notable transformation of terms. Convinced from the get-go that the universe is "progressive," Thomson gradually shifts his own terms until "dissipation" dominates. We know, moreover, that the term carried moral connotations to Thomson long before he adopted it to describe the progress of the physical universe, his father having written to him of the nightly amusements of a certain Glasgow student who had "been indulging in habits of dissipation."[67] Like New York, the universe is indulging in habits of dissipation. It can progress only and inevitably toward entropy—unless, of course, "sources now unknown to us are prepared in the great storehouse of creation."[68]

A Statistical Variation

To the Vril-ya, it is obvious to the barest intelligence that progress in the usual sense is not possible under such a system: "There is no hope that this people, which evidently resembles your own, can improve, because all their notions tend to further deterioration." The specific notion that tends to further deterioration is the maintenance of vast differentials. The "intolerable . . . disparity" that precedes entropic states, such as France during the Revolution, is the result of a specific thermodynamic system in which "a few individuals are heated and swollen to a size above the standard slenderness of the millions." People so distinguished are said to be a mark of progress, the "exceptions to the common littleness of our race [that prove the] magnificent results of [the] system!"[69] But the Vril-ya think otherwise. This is not only because progress so obviously implies dissipation, but also because they understand the drive to social entropy as essentially statistical: "For where societies are large, and competition to have something is the predominant fever, there must be always many losers and few gainers."[70] To the Vril-ya, it is about the distribution of heat characteristic of any many-body system.

This brings us to a new development, a mathematically sophisticated version of energy physics developed in the 1860s and dubbed *statistical mechanics* by James Clerk Maxwell in 1871, the year in which *The Coming Race* was first published. Bulwer-Lytton's extremely timely observation on the statistics of large societies strongly evokes the efforts of Maxwell himself and the subsequent derivations of Austrian physicist Ludwig Boltzmann. Generally referring to the speeds of molecules in a gas, the *Maxwell-Boltzmann distribution* tells us how many particles are likely to be going at each speed. And even though the particles are constantly bouncing off one another and changing speeds, we don't have to track them individually (thank goodness), because the distribution of speeds depends only on the temperature of the gas. (This makes sense if you follow the definitions. The speed tells us the kinetic energy or energy of the motion of each particle. The sum of the kinetic energies, if you recall, is called the thermal energy of the system, the transfer of which is called heat. Their average is called its temperature.) And so statistically fundamental is the Maxwell-Boltzmann distribution, so simply derived from the mathematics of large numbers, that it's not just for gases anymore. Indeed, it never was. For when Maxwell "first obtained the normal distribution for molecular velocities in a gas" in 1859, he got it "directly through the gas-crowd analogy." He took his inspiration from social statistics, "after having become familiar with [Adolphe] Quetelet's curve for social statistics."[71]

Any system made up of a large number of things, having perhaps a few qualities of randomness, mobility, and so on in common with a gas, may evince a Maxwell-Boltzmann distribution in one or other of its internal properties. For us, this means: society is a gas. Or so implies the Vril-ya conviction that "where societies are large, and competition to have something is the predominant fever, there must be always many losers and few gainers."[72] Rereading this sentence with statistics in mind, we recognize at once the usual caveat "in a statistically significant sample" in "where societies are large." The further specification that "competition to have something is the predominant fever" suggests that everyone or everything in the system is driven by the same processes of exchange. Under these conditions, what results is a Maxwell-Boltzmann distribution, in this case, of wealth. In this distribution—what you get when you start with the initial conditions described and run the math—the number of elements above a certain threshold (gainers) will be exponentially small (few) compared to the number of elements (many) below this threshold (losers). A Maxwell-Boltzmann graph of this Vril-ya conviction might look something like that in figure 2.1.

The few above the threshold—those "heated and swollen" above the norm—are simply the statistical variation that indicates the net loss in the system as a whole. This overall loss is apparent in an alternative definition of "Koom-Posh—viz., the government of the ignorant upon the principle of being the most numerous." Difference and contest make up the standard mode of exchange; such a society "place[s] the supreme bliss in the vying with each other in all things . . . for power, for wealth, for eminence of some kind." It is a "system which aims at carrying to the utmost the inequalities and turbulences of mortals."[73]

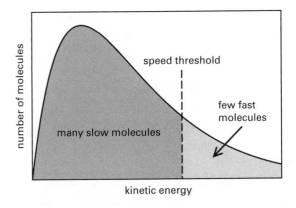

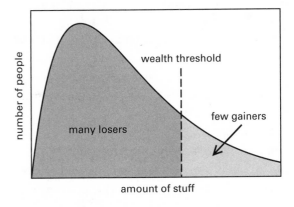

Figure 2.1

The Maxwell-Boltzmann Distribution (Re)Applied to Economics. Among the molecules which make up a gas, there will be many more slow (sometimes referred to as *cold*) than fast (or *hot*) molecules. According to the Vril-ya, a similar distribution occurs among the haves and have-nots in large societies based on competition for wealth.

The Vril-ya, needless to say, strongly disapprove. Holding that "the wisdom of human life [is] to approximate to the serene equality of immortals," such a system simply "[flies] off into the opposite direction."[74] The Vril-ya strive rather for that equilibrated state of "serene equality." Instead of the tensions that drive progress-qua-dissolution, the Vril-ya propose "repose," which "they rank . . . among the chief blessings of life." Take away the "incentives to action which are found in cupidity or ambition" and a human being "rests quiet."[75] The reverse is also true; aim for rest and such incentives are removed. Furthermore, what works for one, works for many. As the Vril-ya reiterate the fantasy of equilibrium, they continue to consider not only the individual, but also the statistical average. Their "society . . . aims at securing for the aggregate the calm and innocent felicity [presumed] to be the lot of beatified immortals." In such a society, "the motives that impel the energies and ambition of individuals in a society of contest and struggle—become dormant or annulled."[76]

Equilibrium, serene equality, calm do not come without cost, however. In removing incentives to action, in annulling contest and struggle, in failing to "impel the energies of individuals," one not only secures peace, but also removes or radically reduces the capacity to get things done. Like a heat engine or a waterwheel, it is difference that drives not only dissolution, but also action or work. Repose as a cultural strategy has advantages and disadvantages. Certainly, it slows progress toward a final energy state. And it is supposed to secure the best living conditions to the aggregate, raising the statistical average. But as an alternative way of thinking about heat death, it may reveal an excess of optimism.

The Problem with Rest

The Vril-ya are not insensitive to this implication of their drive toward equilibrium. Where, after all, is the line between restful and at rest? Still "irrationally valuing equity over efficiency" (an intentionally ironic, but unintentionally thermodynamic phrase I happened to spot on a T-shirt),[77] the Vril-ya do not put the problem as explicitly as does Spencer, for whom it represents the chief challenge to his evolutionary optimism: "But the fact it chiefly concerts us to note is that this process [of evolution, broadly construed] must go on bringing things ever nearer to complete rest."[78] Rest thus serves as a synonym for heat death. Indeed, the Vril-ya's anxiety that an advanced culture must necessarily be more entropic reveals itself even in the diffuse light that they maintain throughout their realms. But eventually, that diffusion attaches to a certain ambivalence about the advanced state of their culture. Note that "they maintain, both in the streets and in the surrounding country, to the limits of their territory, the same degree of light at all hours. . . . their lights are never wholly extinguished." The reason is this: "They have a great horror of perfect darkness."[79] The steady state of light thus seals the advanced state of equilibration among the Vril-ya even as it heralds the inevitable next step. Undoubtedly horrified of the dark because they live underground, the Vril-ya have a horror of darkness that signifies more. For the final equilibrium—when all light is fully diffused, all heat at uniform temperature, all energy beyond its usefulness—implies the end of all things. "Perfect darkness" reiterates this ambivalence, signaling at once the state toward which this advanced culture tends and the point at which this advanced culture ends.[80]

The narrator, not surprisingly, prefers sunshine. Impressed by the "ingenuity and disregard of expense" with which the Vril-ya illuminate "the regions unpenetrated by the rays of the sun," he is convinced that it is ignorance only that reconciles them to their lot. As he puts it, "I could not conceive how any who had once beheld the orbs of heaven could compare to their lustre the artificial lights invented by the necessities of man."[81] He is, as it happens, unable to see the careful attention to expense that underlies the lighting choice of the Vril-ya. Thermodynamically speaking, the narrator prefers a light source that is much more wasteful. He can afford to, because he enjoys the benefits of an energy source that has dissipated relatively little. He has the sun.

The sun, moreover, illuminates more than just the surface of the earth. The narrator's preference for light that penetrates, rather than light that diffuses, reveals his politics as well. As his ambitions rise, he reflects on the imperial possibilities of Vril. "Of course," he muses, "one would not go to war with the neighbouring nations as well armed as one's own subjects"— that is, with nations of the Vril-ya. Like mechanical work, the work of empire can be driven only by difference. But there are plenty of nations, even below ground, who haven't got vril. Significantly, these "[resemble] in their democratic institutions, [his] American countrymen," but our narrator is content to take them over anyway: "One might invade them without offence to the vril nations, our allies, appropriate their territories, extending, perhaps, to the most distant regions of the nether earth, and thus rule over an empire in which the sun never sets." Deploying this classic descriptor of the British empire, our narrator suggests how the metaphor of the sun asserts, in both imperial and energetic terms, that power resides in an identifiable location. In contrast to the diffuse light shared by all the Vril-ya nations, the powerfully localized light source functions as the emblem of localized imperial power, neither diffused nor equilibrated nor shared.[82] Of course, the narrator "[forgets], in [his] enthusiasm, that over those regions there was no sun to set,"[83] pointedly denying the temporary nature of the imperial power he enjoys (or hopes to) as well as the inevitability that the sun will set (in the short term) and diffuse away (in the long). Bulwer-Lytton thus gives us an early example of what will become a popular metaphorical association between heat death and national decline, as in Max Nordau's "Dusk of the Nations, in which all suns and all stars are gradually waning, and mankind with all its institutions and creations is perishing in the midst of a dying world."[84]

Taking a less grim view than Nordau will, our narrator's pointed denial of the approaching Dusk is manifest in his discomfort with Vril-ya culture and in his preference for the fast and furious lifestyle possible only with a localized, wasteful, and relatively untapped energy source. Not particularly savvy about the energetic implications of the steady state maintained by the Vril-ya, he is simply bored. For our narrator, social equality implies the end of art, work, progress. Acknowledging that the "social state of the Vril-ya [had] contrived to unite and to harmonize into one system" nearly all of the Utopian ideals of the upper world, and further that "Equality . . . was not a name; it was a reality," he nonetheless wishes it all away in favor of "change, even to winter, or storm, or darkness." What once struck him as "so holy a contrast to the contentions, the passions, the vices of the upper world, now [oppress him] with a sense of dulness and monotony."[85] His aspirations prove pointedly "restless," and he favors difference or change even if these drive degeneration. Those things that signal heat death to the energetically aware—winter (cold), storm (disorder), and darkness—are to our narrator acceptable alternatives to the dullness and monotony of maintaining a steady state. Either he doesn't see it coming or he doesn't care.

Spencer, on the other hand, sees it coming from a long way away. As we follow the implications of thermodynamic equilibrium to their logical conclusion, we eventually come to a full stop. The "final question," which his readers have probably noticed, is this: If evolution, broadly construed, is that increase in complexity "that is incidental to the universal process of equilibration, and if equilibration must end in complete rest, what is the fate towards which all things tend?" *All things*, of course, include the solar system, which is "slowly dissipating its energies . . . the Sun is losing his heat." And pretty soon—in "millions of years"—we will feel it. "Geologic and meteorologic processes" will slow; "the quantity of vegetable and animal life" will decrease. As Spencer makes this final move to unify, things hit closer and closer to home: "If Man and Society are similarly dependent on this supply of energy which is gradually coming to an end; are we not manifestly processing towards omnipresent death?"[86]

Spencer reiterates his conviction that the laws of thermodynamics drive all systems, even as he is forced to concede that thermodynamically speaking, "equilibration was death." He makes this concession in an 1858 letter to his friend John Tyndall, who had apparently explained that "when equilibrium was reached, life must cease."[87] Tyndall is right. When the whole system has come to uniform temperature, we can't drive a heat engine. When all the water is at the same level, we can't turn a waterwheel. When the sun goes out, we die. All processes meteorological, geological, social, and personal must come to an end. Spencer fears not only the extinction of the sun, but also the more immediate consequences of pursuing equilibrium. His same understanding of evolution as increasing complexity applies to changes in the law, as "each new law, beginning as a vague proposition, is . . . elaborated into specific clauses." But the increase in complexity cannot elude the consequences of equilibration, for as we apply the law, its purposes become "more precisely formulated" and its "action more restricted . . . until at last decay follows a fixity which admits of no adaptation to new conditions."[88]

Both Spencer and Bulwer-Lytton identify the absence of change with final rest or entropy. Decay follows fixity. The question of complexity divides Spencer's thinking from that of the Vril-ya, however. Where the Vril-ya prefer the simplification in government implied by philosophical unity, Spencer wants to see complexity as more evolved, but cannot: "Thus far no reason has been given why there should not ordinarily arise a vague chaotic heterogeneity, in place of that orderly heterogeneity displayed in Evolution." In fact, as much as he wants to see an increase in complexity as progress, he cannot rule out a chaotic heterogeneity. This possibility, of course, has political applications. They manifest themselves in the dissolution of rank when "all magisterial and official powers, all class distinctions, all industrial differences, cease: organized society lapses into an unorganized aggregate of social units."[89] Spencer's fear requires us to reexamine the Vril-ya's proclaimed preference for the dissolution of rank. Their proverbial "No happiness without order, no order without authority, no authority without unity," which hitherto bespoke the diffusion of power in the Vril-ya state, now reminds us that authority remains. Their equality is carefully

qualified: "Not that all are equals in the extent of their possessions or the size and luxury of their habitations: but there being
no difference of rank or position between the grades of wealth
or the choice of occupations, each pursues his own inclinations
without creating envy or vying."[90] Wealth is not evenly distrib-
uted and important distinctions—an orderly heterogeneity—
remain. The "serene equality" of the Vril-ya suggests that their
equilibrium is neither complete nor final.

Equilibrium on the Move

Fully cognizant that complete equilibrium means the end of all
things, both Spencer and the Vril-ya devise a carefully qualified
equilibrium instead. Spencer's reminds us that the laws of thermodynamics obtain only within a closed system—even if that
system is as large as the universe. Energy is always conserved;
entropy, always increasing, as long as nothing is getting in or
out. Only then must Spencer's equilibration bring us to complete rest. Thus we find a loophole in the law of entropy. The
system may be opened: "Energy ever in course of dissipation, is
ever renewed from without: the rises and falls in the supply
being balanced by rises and falls in the expenditure." Spencer's
easy use of *supply* and *expenditure* reiterates the pervasive association of energy and economics, even as he grounds his social and
economic notions in the practical developments of engineering.
We can have a kind of equilibration, what he calls the "third
order," in systems "which continually receive as much energy
as they expend. The steam-engine . . . supplies an example." The
relation between engines and economies, nations and the universe, is at once linguistic, metaphorical, and physical. All are
governed by the same principles; the same kinds of equilibra-

tion may be found at all scales because all derive from the first principle, the persistence of force. And in all cases, the loophole obtains. In this "penultimate stage" (it turns out there are four orders of equilibration), we may achieve what Spencer calls a "moving equilibrium" and what chemists will call a "steady state"—maintaining all the advantages of equilibrium without the nasty end-of-all-things part.[91]

Of course there is still *a* nasty part. As a similar loophole gradually reveals itself in the carefully equilibrated social order of the Vril-ya, we find that equity within the system depends on disparity without. Things may seem fairly well equilibrated among the "vast general family" of the Vril-ya. They share language, law, and customs. They intermarry. And they all know about vril. And where happiness, so well diffused, has defined civilization, the word *A-Vril* now serves that function; "Vril-ya, signifying 'The Civilized Nations,' was the common name by which the communities employing the uses of vril distinguished themselves." Energy unites both notions and nations. But it also divides, because these same energy principles place a limit on equilibration. There must be difference if work is to proceed, so it becomes necessary to define an *outside*. Thus the Vril-ya "distinguish themselves *from* such of the Ana as were yet in a state of barbarism."[92] The attitude Zee exemplifies toward the savage Other is therefore not incidental but essential to maintaining the qualified equilibrium that exists among the nations of the Vril-ya.

The dependence on a differential is not only the energetic principle that drives evolution and, of course, the steam engine, but also that which will secure global domination: "Since in the competition a vast number must perish, nature selects for preservation only the strongest specimens," and Vril-ya legend holds that they "are destined to return to the upper world, and supplant all the inferior races now existing therein."[93] As Bruce Clarke, a scholar of literature and science, observes, "This vision of the future ruin of the present human chaos is the novel's final metonymy of thermodynamic entropy."[94] It is this and more, entangled as it is with the discursive evocation of evolution, empire, and the conditions required to produce order. Meanwhile, "survival of the fittest" is inflected by Thermo-Poetics; the phrase now evokes the disparity in the distribution of energy or power required to do work. Evolutionary and political competition, moreover, exhibit the usual statistical distribution wherein a few individuals benefit at the expense of the many and, indeed, the mean. Vast numbers perish; only a few cross the threshold of fitness. And lest we fool ourselves that such principles apply only in the early processes of history, we see that even the slow progress of the Vril-ya tends toward net deterioration as they make plans to open the system and supplant the vast numbers inhabiting the upper world. They will, in the process, once again assert the superiority of their (knowledge of) energy, with "powers surpassing our most disciplined modes of force." And even their virtues will prove "antagonistic in proportion as our civilization advances." No wonder our narrator ends with a prayer "that ages may yet elapse before there emerge into sunlight our inevitable destroyers."[95] Again, force divides, as the Vril-ya acquire a new energy source, expanding their order at the cost of the inevitable destruction that reminds us that in the larger system, entropy must always increase.

3   The Reign of Force

Thy reign, O force! is over.
—James Clerk Maxwell

Facts in the Making

Characterized by tension, even paradox, Victorian thermodynamics exhibits some of the protean quality of energy itself.[1]
Those who write about it find it transforms easily, providing
metaphor or even physical justification for pet theories of all
kinds. And vice versa: when thermodynamics is thus used, that
usage solidifies the status of its statements as facts rather than
artifacts. As this chapter extends *ThermoPoetics* to a different
kind of literature—the poetry and popularizations of Victorian
physicists—we will consider how these participate in the processes of transformation that make statements into facts, those
indisputable assertions "devoid of any trace of ownership, construction, time and place." To do so, it helps to introduce some
of the ideas of Bruno Latour, whose book *Science in Action* investigates what he calls "the most important and the least studied
of all rhetorical vehicles: the scientific article."[2] He shows how
statements made in these articles are "transform[ed] . . . in the

direction of fact or of fiction," depending on what "listeners" (readers, other scientists, writers of subsequent articles, and so on) do with them. He is referring to science in the making, to the processes by which scientific ideas come to be accepted or rejected. A claim is made; subsequent readers, writers, and speakers help it along—using it, confirming it, building on it, sending it "downstream," as Latour puts it, in the direction of fact. Or the same claim is made, and others question it, elaborate doubts, ask questions like "Who said it?" "Based on what?" and "Why?"—sending it back "upstream" to the circumstances of its production and calling into question its fact*ness*, making it an artifact.[3] Such fact-making processes are, of course, not limited to scientific articles. And in this chapter, I propose to explore what happens to scientific statements—terms, ideas—as they are spread about. When a scientist such as James Clerk Maxwell writes poetry, he sends some notions upstream and some down. When a popular essay explains the distinction between kinetic and potential energy, it makes both of these more real. And when energy concepts are explained through political metaphor, the effect is mutual—the political and the physical may well sink or swim together.

Forceful Language

The protean quality of the laws of thermodynamics intensifies while the language of thermodynamics is itself undergoing transformation. Though this is true of so many words associated with energy, and very true of *energy* itself, P. G. Tait's 1876 lecture on "Force"[4] identifies this one as a particularly "abused and misunderstood term." Acknowledging that "in popular language there is no particular objection to multiple meanings for the same word," even that the existence of multiple meanings is "one of the most fertile sources of really good puns,"[5] Tait struggles to distinguish scientific language from poetic, pu(g)nacious, and colloquial uses. He seeks to establish a singular scientific meaning for *force*—one that, moreover, subordinates it to the more important term *energy*.[6] However, when his friend James Clerk Maxwell—a physicist of much wider and more enduring fame than Tait—writes a poem in honor of Tait's lecture, he seems to enjoy the multiple meanings of *force*. In fact, he seems set on taking his place among Tait's great pun makers, "Coleridge, Hood, Hook, or Barham."[7]

Sharing with Tait and others the sentiment that force is an inexact term, which must give way to a reduced role in an energy-centric worldview—indeed having written to Faraday to urge him to clarify his use of the term (see the previous chapter)—Maxwell nonetheless enjoys its lingering ambiguity. His poem "Report on Tait's Lecture on Force" suggests that the role of force in the world is over in more ways than one:

Force, then, is Force, but mark you! Not a thing,
        Only a Vector;
Thy barbèd arrows now have lost their sting,
        Impotent spectre!
Thy reign, O force! is over. Now no more
        Heed we thine action;
Repulsion leaves us where we were before,
        So does attraction.[8]

The antiforce sentiment expressed here runs throughout Maxwell's poetry; his poems develop a pretty consistent picture of the wrongness of force. Characteristic of the old reign, force is dispelled by the new physics, which reduces it to a mathematical construct: "only a vector," rather than a substantive threat or physical entity, an actual "thing" that can do harm and that "we" need heed. Maxwell puns even on the graphic arrows used to represent vectors. In this way, "arrows" lose their sting, becoming not the weapons of an oppressive and primitive old regime, but the mere symbolic representation of a quantity with direction. Maxwell's punning thus makes a clear effort to send force, as a physical notion, back upstream. As we will see as we look at more of his poetry, *force* becomes a highly suspect term, a term with a history, an artifact.

But a pun—by pointing to the different meanings of a word— can make a double move, sending some meanings upstream and some down. Milking this pun for all it's worth, Maxwell suggests that force is not only passé in physics, but also no way to run an empire. While *The Coming Race*, as we have seen, suggests that energy physics motivates imperial expansion as the best way to avoid national heat death, this conclusion hardly exhausts the possibilities of thermodynamics for politics. Apparently delighted that the new physics also implies a kinder, gentler regime, Maxwell proceeds to even greater heights of postrevolutionary hyperbole. Not only are we now impervious, but "the universe is free from pole to pole,/Free from all forces." Of course, Maxwell picks a rather strange moment to declare the universe free. We may find unintentional irony in the timing of his declaration—1876, the year Victoria crowns herself Empress of India and the British empire may be said to be at its height (or rather its width). But Maxwell's discourse on force at best glosses anything potentially problematic in the contemporaneous workings of the British empire, at worse, declares it forceless. "Force" becomes an oppressive regime in itself, detached from specificity of nation or history. If it remains historically specific, it is in its association with relatively benign old physics.

Maxwell is not alone, moreover, in the impulse toward forceful hyperbole or in projecting the discourse of empire, nation, war onto the contention over scientific theories. We find a similar kind of language, for example, in an anonymous review of Joule's papers that declares that none of the greats of physical science "more thoroughly shattered a malignant and dangerous heresy, than did Joule, when he overthrew the baleful giant FORCE, and firmly established, by lawful means, the beneficent rule of the rightful monarch, ENERGY!"[9] Already "a theme for the Poet of the Future," this heroic account of the progress of science—especially the glorious revolution of energy, "more swift and more tremendous than ever befell a nation"—establishes Joule in a line of such heroes, among them Maxwell and Newton. What is elided, moreover, is not only the literal revolutions that have befallen or might befall nations, but also the fact that Newton himself was the great proponent of "FORCE." Newton's three laws tell us how to calculate forces and have embroiled generations of students in the drawing of force diagrams to determine what forces are acting on what and with what results. If indeed he is overthrowing the "baleful giant FORCE," Joule must necessarily be altering the status of Newton. Of this, however, Anon seems unaware. In spite of the implications wrought by the association of "FORCE" with heresy, "Newton [who] annihilated the Cartesian vortices" figures here as the hero of old, whose legacy Joule most properly assumes. The progress of science marches onward and upward as revolution after revolution establishes the rule of right over might. Such moves are more or less what we expect from a publicist for the science of energy and more particularly for Joule, who figures as the rightful heir to a long line of scientific heroes. Moreover, in exalting "ENERGY" as the "rightful monarch," this writer effects an elision—or a displacement—similar to Maxwell's. The outcome of revolution is still monarchy, though revolution shifts to that which occurs within science.

Maxwell's poems, however, reveal a set of more complex relations to and among Newton, science, and politics. On the one hand, they evoke some of the philosophical problems underlying the program of mechanical explanation that drove nineteenth-century physics, including questions of the place of mathematics and analogy in science. On the other hand, they suggest the difficulties besetting a community of scientists eschewing force while claiming a Newtonian pedigree. The ease of slipping into criticism of the paterfamilias leads to considerable work to divide Newton's good children from his bad, work that necessarily brings force back into the spotlight.

Only a Vector

Maxwell's poetry casts doubt on the legacy of Newton in two ways—one mathematical, the other political. That "Not a thing/ Only a vector" crack evokes the controversy over the status of mechanical explanation itself. As force comes under new scrutiny, so too does its status as a natural independent entity (a thing) as opposed to a mathematical or metaphorical construct (only a vector). Though generally very subtle on the reality status of mathematical analogies, Maxwell here evinces no uncertainty as he answers Tait's question of "whether there is such a *thing* as Force, or not."[10] Where Tait himself acknowledges the thingness of some force, Maxwell's "Report" says emphatically *not*. At least, its thingness cannot touch us, nor does it require our attention. Maxwell seems to send force back upstream, without a paddle. But the thing/vector dichotomy points us to a larger question even than that of force.

The overarching, unifying project of nineteenth-century physics has been productively described as the widespread attempt to explain everything mechanically. At the same time it becomes increasingly clear that we don't know what mechanical explanation means. Can we—should we—make claims that our explanations are actually descriptions? For example, when we make sense of heat by describing it as a "mode of motion" or even as the "energy of these motions,"[11] what are we actually talking about? How far are we willing to go? Are we claiming that we know what's moving and how? Can we trace or predict the movement of actual stuff . . . molecules, say? Or is it sufficient to say that stuff is moving in ways analogous, even transferable, to other stuff? And what does it mean when we do the math? Is that a description? An analogy?

Maxwell is famous for his use of mathematical analogy. Educated at Cambridge, he joined a group of physicists increasingly committed to incorporating sophisticated mathematics in their explanations and predictions of physical phenomena.[12] But when he came up with the idea of describing the behavior of electricity and magnetism using the mathematics of fluid flow, what was he after? Was he claiming that electricity *is* a fluid? Or is it just a metaphor? Of course, he claimed nothing so simple as the former, nor could he (or we) dismiss the truth benefits of the latter. Thus while carefully qualifying how much descriptive value can be ascribed to various mechanical models, he nonetheless sees value in positing possible explanations without claiming an actual "connexion existing in nature."[13] He admires what he calls Thomson's "allegorical representation" of electrostatics, but also insists that "any mathematical formalism must keep the physical problem clearly in focus"[14]—not that keeping the physical problem "in focus" actually clarifies what's going on here. And his statistical methods—the mathematical formalism he applies to molecules and gases—point again to the limits of what we can keep in view. We cannot follow the motions of individual molecules; we can only speak to the behavior of the aggregate.

Poetry, as it happens, is a good place to work out all these ambivalent feelings. Maxwell's "only a vector" masks a much more complex relation to the value of mathematical expression. The vector, devoid of force, can no longer harm. But a much earlier poem, "A Vision: *Of a Wrangler, of a University, of Pedantry, and of Philosophy*" (1852), expresses a concern that mathematics may in fact do harm.[15] Here he suggests that nature is belittled by scientific endeavors to describe it through mathematical means: "Such the eyes, through which all Nature/Seems reduced to meaner stature." Asking, "Is our algebra the measure/ Of that unexhausted treasure," he suggests that mathematical description proves generally inadequate, and that, in fact, it is presumptuous on our part to stop with such description, that "Man forgets his station/If he stops when that is done." Perhaps keeping the physical problem in focus addresses this inadequacy. Perhaps poetry is required.

But mathematical construction figures elsewhere as poetry and even as music. Maxwell's poem titled "A Problem in Dynamics" satires the differential calculus (yes, really) involved in determining the forces acting on an "inextensible heavy chain."[16] We may even speculate that the convolutions of such calculations were what gave the twenty-two-year-old Maxwell an early distaste for the conservation of force. In another poem, he plays on the etymological connection he makes in naming the mathematical symbol ∇ after the "Assyrian harp" of similar shape, the Nabla.[17] This naming reinforces the claim in the poem that what mathematics provides is less a natural description than a human construct, answering a human need. Scientific experimentation (itself, Maxwell suggests elsewhere in the poem, a process involving a good deal of singing and dancing) provides us with "transient facts . . . fugitive impressions [that must] be transformed by mental acts,/To permanent possessions." To effect this acquisition, itself subtly suggestive of the colonization of nature, we summon up our "fancy scientific" until thought combines with "sights and sounds [to] become of truth prolific."[18]

Newton's Heirs

Of course, not everyone can produce poetry out of calculus (which is not to say we shouldn't try). Arguably, Maxwell has done so in an enduring way in his equations of electricity and magnetism. His poems suggest that Tait can too; biographers Campbell and Garnett identify Tait as the titular "Chief Musician upon Nabla."[19] And in the "Report" Maxwell shows us how it's done, as "Tait writes in lucid symbols clear/ One small equation;/And Force becomes of Energy a mere/ Space-variation." One small equation not only frees the universe from forces, but also frees us from the convolutions of force calculations.[20] That's the way to do Newton, Maxwell seems to say. Retaining force as a space variation of energy enables us to keep Newton without bowing to the more malevolent uses of force. It may be that "Action and Reaction now are gone," but we needn't sacrifice Newton into the bargain. Indeed, Tait's original lecture goes one better, suggesting the ways the North British School more broadly managed to make Newton into the father of Victorian energy physics. The lecture claims to "translate Newton's words (without alteration of their meaning) into the language of modern science."[21] Tait further does his best to credit Newton with (almost) discovering the first law of thermodynamics, with telling us that "*energy is indestructible*—it is changed from one form to another, and so on, but never altered in quantity." Of course, as Tait himself acknowledges, there is an important loose end: "To make this beautiful statement complete, all that is requisite is to know *what becomes of work done against friction.*" It is a regrettable oversight that keeps Newton himself from noticing the relation between friction and heat; Newton "seems to have forgotten that savage men have long since been in the habit of making it [friction] whenever they wished to procure fire."[22] Fortunately, Tait appears, translating Newton's laws into the language of modern science and the language of force into the language of energy. His move also construes energy transformation as a kind of translation; both language and energy are supposedly changed from one form to another without significant alteration, whether of meaning or quantity. Considerable work is involved in construing Newton this way, but it proves well worth the effort.[23]

Even as Maxwell celebrates and elevates "brave Tait" he sends "force" further upstream by vilifying those who not only use, but abuse "force." In Maxwell's "Report," Tait figures as a level-headed scientist and brave revolutionary hero, who knows "so well the way/Forces to scatter/Calmly await[s] the slow but sure decay,/Even of Matter." But his calm heroism in the face of universal heat death stands out in stark contrast to those pretenders to Newton's legacy. Tait, the fisher of physicists, comes "with his plummet and his line,/Quick to detect . . . Old bosh new dressed in what [is called] a fine/Popular lecture." Ironically evoking a line from Pope's *Essay on Criticism*—"True wit is Nature to advantage dress'd"—Maxwell accuses these "British Asses" of the cowardice of incoherence. Such lecturers stay "safe by evasion." Their "statements baffle all attacks"; their "definitions, like a nose of wax,/Suit each occasion." But foppery slides into deception as these wranglers make of "that small word 'Force,' . . . a barber's block,/Ready to put on/Meanings most strange and various." Thus they shock the true "Pupils of Newton" as well as taking in the "lordlings and ladies" who would trust them. Maxwell accordingly reiterates a move we have seen within *The Coming Race*. Force serves again to divide the good scientists from the bad, the true heirs from the pretenders. In doing so, moreover, he evokes the tensions of race and nation that figure prominently in Bulwer-Lytton and in the broader discourse on force. Representing wranglers as a "race" that abuses force, Maxwell promises that their arrows can no longer wound.

But in constituting forces as "no more [than] symbols of disgrace" that cling to those who still peddle outdated science,[24] Maxwell elides a larger issue. For forces are hard to shake in a Britain evincing considerable guilt over its imperial past, guilt that the progress of empire has been less than forceless. That science wars cannot be fully disconnected from international tensions becomes evident in the "Report," as we find that the continued misuse of force represents an attempt to cling to methods that are at once violent, ancient, and foreign:

Ancient and foreign ignorance they throw
        Into the bargain;
The shade of Leibnitz mutters from below
        Horrible jargon.
The phrases of the last century in this
        Linger to play tricks—
Vis Viva and *Vis Mortua* and *Vis*
        Acceleratrix:—

Clinging to the old litany of *Vis*'s shows oneself to be not only stuck in the eighteenth century, but also decidedly unpatriotic. This is a good example of how poetry functions as literature's microchip. There is a lot of information in a very small space. In fact, so much converges in Maxwell's accusation that it is hard to pin down, but here goes. Tait's lecture systematically demonstrates that most of what were called *Vis*'s cannot properly be called forces. Some do not even merit the status of things, though Vis Acceleratrix does, "being really no force at all, but another name for . . . Acceleration."[25] Tait's discussion of Vis Inertia evokes the inevitability of puns, though Tait tries to keep them personal rather than national, insisting that it "implies, not revolutionary activity, but dogged perseverance."[26] But no amount of disclaiming against multiple meanings can keep us from reading "revolutionary activity" as more than rotational motion, or dogged perseverance as a characteristic only of physical bodies. Vis Viva fares best, Tait proclaiming that it is indeed a thing, "*Kinetic* Energy," though not the thing we thought it was, "*Living Force.*"[27] And Vis Mortua, though once used to refer to a force doing no active work, seems here to be Maxwell playing on words, chuckling as he declares once more that force is or ought to be dead.

The *Vis*'s, moreover, evoke the international contention that beset Newton in his day and lingers in Maxwell's. Leibniz figures not only as Newton's chief competitor in the invention of calculus, but also as the initiator of the *"Vis Viva* Controversy"—a retrospective look at which demonstrates the confusion that arises in claiming force as a conserved quantity before reaching consensus on what force means.[28] According to Tait, "a good deal of the confusion about Force is due to Leibniz,"[29] who was, of course, German. And his ghost lives on in the claims of German scientists to nineteenth-century advances in thermodynamics. Maxwell was willing to give Clausius his due. But John Tyndall's efforts (about which Tait was vociferous) to elevate J. R. Mayer's claims at the expense of Davy's, Joule's, and even Thomson's would not have sat well at all. Alternatively, "foreign ignorance" might allude to "continental action-at-a-distance theories," especially those of Wilhelm Weber, whom Maxwell criticized for violating the "conservation of energy" in his theories of electricity.[30] Or perhaps "foreign ignorance" signals his frustration at the incomprehensibility of Boltzmann, with whom Maxwell's name is now forever coupled.[31] "Maxwell's disquiet about German thermodynamics" in turn evokes the problem of mechanical explanation. But in the "Report," the prime offenders are not the German scientists, no matter how much they may strive to "reduce the second law to mechanics,"[32] but those "British Asses" who would return us, one way or another, to a universe of force.

Stewart and Tait

In spite of his heroics in the campaign against force, Tait too finds force, imperial and physical, hard to avoid. He and his coauthor Balfour Stewart come to represent science as something between a foreign invasion and a natural disaster. In their

very popular book *The Unseen Universe*, they take on the task of
the romantic hero we first met in Maxwell, setting out to protect
the innocent against certain practitioners of science, though
certainly not against Newton or science itself. Though anxious
to reconcile the new physics with a spiritual understanding of
the universe, the authors nonetheless express their sympathy
with those who fear the advance of science. Blithely mixing
their metaphors, Stewart and Tait suggest not only the potentially destructive power of the rapid progress of knowledge, but
also its naturalness, even its inevitability: "Like a wave swelling
as it advances shoreward, this progress has violently transformed
whole regions of thought, while it has repeatedly invaded
others." These latter regions, we soon understand, are "occupied
by the followers of Christianity," an unsuspecting and relatively
compliant population who fear not just a tidal wave, but another
deluge in "the recent great floods of intellectual energy . . . which
have repeatedly invaded the region." The energy in question is
at once intellectual and physical; the recent flood of intellectual
energy most interesting to these authors is the new science of
energy itself. As proponents of the science of energy determined
to demonstrate its fundamental compatibility with both a spiritual life and an afterlife, Stewart and Tait position themselves
as intellectual missionaries, seeking to answer "an almost
despairing outcry from many of the inhabitants of the Christian
region" and "to reassure these somewhat over-timid people."[33]

In this way, Stewart and Tait reflect and expand on a fundamental belief held by many of the early intellects of energy,
including Thomson, Joule, and Watt, all of whom found or
created the new science to be fully compatible with certain
liberal interpretations of Christian scripture. Like Maxwell, as
well as compatible with the more partisan goals of the North

British school to which Tait has especially strong ties, *The Unseen Universe* understands some scientists as the rightful heirs of Newton, some intellectual thought as advancing in a kinder and gentler way. Undoubtedly doing their part to discredit Tyndall,[34] they thus distinguish between the "splendid example shown by intellectual giants like Newton and Faraday" and the "materialistic statements now-a-days freely made (often professedly in the name of science)" that not surprisingly trigger such alarm among the timid, indigenous population of Christians.[35]

But where science is done right, Stewart and Tait suggest, its right to advance is beyond question. The principles of science are the rightful and apparently inevitable successors to spiritual modes of thought throughout history, including that of the ancient Egyptians, Hebrews, Greeks and Romans, the Eastern Aryans, the disciples of Christ, of Mohammed, and more recently Swedenborg. Unreflective about their own intellectual appropriation of ancient and frequently Eastern peoples, they suggest that the conservation of energy is fully compatible with a belief in the immortality of the soul, actually giving us the first statement of conservation in the context of ancient Egyptian belief: "Dissolution, according to them, is only the cause of reproduction—nothing perishes which has once existed, and things which appear to be destroyed only change their natures and pass into another form."[36] Thus, in their attempt to popularize the new science of energy, Stewart and Tait tread dangerously close to what a detractor could call "old bosh new dressed" or perhaps "ancient and foreign ignorance." But in *The Unseen Universe*, their efforts to claim the legacy of Newton seem untroubled by such concerns, and their desire to establish the compatibility of science with religion far outweighs any anxiety regarding what may come to look like pseudoscience or half-baked philosophy. Notably for our purposes here, energy figures as the proper heir to ancient as well as modern thought. As in Maxwell, it opposes itself to a violent and oppressive regime, especially that of the scientific naturalists. Energy may figure either as that which violently overthrows an old and oppressive regime or that which marks the next stage of progress in the development of various regions (of thought). In either case, whether what went before is violently overthrown or gently converted to new and better ways of thinking, energy appears on the side of what is both good and, with an increasingly resonant pun, conservative: the rightful monarch, pedigreed, the heir or supplanter of Newton, or the next great empire modeled on its ancient and noble predecessors. It is thus through the efforts of many people that the object *energy* is given "the contours that ... provide assent."[37] Throughout England, scientists and Christians alike can embrace the new principles, confident that they fit with what was already known or hoped about people and politics as well as physics. Apparently, one can build a better steamship without overmuch rocking the boat.

Carnot

Outside of England, on the other hand, there are those who wouldn't mind shaking things up a bit more. In correlating, for better or worse, the advance of energy with that of empire, the writers discussed above develop a suggestive and effective metaphor for popularizing, legitimating, establishing the new physics. But not from scratch. They seem rather to channel into metaphor what has already seen a more literal life in the discourse of energy. They prove to be heirs of pre- and postrevolutionary

France in general, Sadi Carnot in particular. Indeed William
Thomson was instrumental in reviving Carnot's *Réflexions sur la
Puissance Motrice du Feu* for broad and especially English consumption. In Carnot's "On the Motive Power of Heat" (1824),
the associations of energy and empire are established early on
in practical as well as metaphorical terms. Carnot was an engineer participating in the technological growth of postrevolutionary France.[38] Because of his father's role as an influential
general and engineer, as well as his own fluctuating military-
political career, Carnot had firsthand experience with the rise
and fall of French empire under Napoléon.[39] As a result, his
central concern is to build a better steam engine. Moreover, the
driving force, one might say, of his scientific inquiry is explicitly
the progress of nation.

For Carnot, the steam engine embodies not only the practical
values of Enlightenment science, but also the distinction
between civilization and savagery. It is the key to industry,
empire, and progress. It "has made it possible to traverse savage
regions where before we could scarcely penetrate. It has enabled
us to carry the fruits of civilization over portions of the globe
where they would else have been wanting for years." Technology serves as both the carrier and marker of civilization. The
current state of steam engineering highlights the rapid progress
that civilization has made: "There is almost as great a distance
between the first apparatus in which the expansive force of
steam was displayed and the existing machine, as between the
first raft that man ever made and the modern vessel." The stuff
of science fiction and imperial fantasy, steam navigation thus
converts time into space and shrinks the earth, "bring[ing]
nearer together the most distant nations." And in what
some might construe as global unification, others, as global

domination, it "tends to unite the nations of the earth as inhabitants of one country."[40]

And though it seems he might like to, Carnot cannot refuse to England "the honor of discovery [which] belongs to the nation in which it has acquired its growth and all its developments." This is an interesting, almost Latourian, definition of discovery, which equates it with growth and development.[41] Nearly synonymous with invention, discovery suggests not the simple uncovering of something always/already in nature and waiting for science to stumble on. It is, rather, marked by systematic development and is driven by desire, such that "it is natural that an invention should have its birth and especially be developed, be perfected, in that place where its want is most strongly felt." But for all of his admiration for this English discovery, Carnot can't resist interjecting a momentary fantasy of British decline, musing that "to take away today from England her steam-engines would be to take away at the same time her coal and iron . . . to dry up all her sources of wealth." It would be—the daydream continues—"to annihilate that colossal power." And while Carnot can also imagine "the destruction of her navy, which she considers her strongest defense," he concludes that this "would perhaps be less fatal" than ruining for England "all on which her prosperity depends."[42]

This somewhat wistful speculation leads to no more detailed plans for the undoing of British empire. It does, however, serve to introduce lasting resonances into the language of steam engineering that eventually come to characterize the discourse on energy. When shortly afterward Carnot observes that "the question has often been raised whether the motive power of heat is unbounded,"[43] he seeks explicitly to fill a gap in the state of the art, to seize a scientific opportunity: for all their practical advances, the theory of steam engines is little understood. But following fast on his speculations regarding the "colossal power" of the British, *power* continues to resonate with imperial as well as engineering significance, and the reader wonders whether the power of the British is indeed unbounded. *Expansive* also retains some of this double significance, and while we eventually focus on the capacity of steam to do mechanical work, there is an important transitional moment wherein the "expansive force of steam" evokes at once both its mechanical and its imperial effects.

Stewart and Lockyer's Account

Of course, we can hardly discuss empire without thinking about economics. Victorians, it seems, found it irresistible to compare energy and economy. We see such comparisons all over the place, from the most casual metaphor dropped in popularizations to the most extensive naturalizations of political economy (that is, in work intended to represent the current state of affairs as the natural or correct state of affairs). One writer opens his account of "A New Theory of the Sun" by suggesting that we traditionally think of the sun as "the great almoner, pouring forth incessantly his boundless wealth of heat"; the new theory may then be regarded as "a first attempt to open for the sun a creditor and debtor account."[44] Elsewhere, we consider the sun's "amazing fund of vitality."[45] Slightly less fanciful is talk about the sun's "store of energy ... garnered up during long past ages."[46] Such uses are not limited to popular or anonymous writers. Tennyson's astronomer friend Norman Lockyer refers to "the treasure houses wherein are kept the secrets of the sun."[47] Even William Thomson makes a well-known reference to "the great storehouse of creation."[48] And it is positively routine to discuss the expenditure of power or to discuss the amount of energy spent to accomplish some amount of work. This particular metaphorical trend, moreover, seems tightly connected to the economic and national concerns driving the science of energy. With Carnot, we have seen these concerns at or near the inception of thermodynamics. By the end of the century, it is a commonplace that the "cheapness of power" is an all-important factor in industrial and national development: "In England, the secret of our industrial greatness and, in a measure, the cause of our commercial supremacy, may be revealed in our unrivalled facilities for the production of . . . heat-power."[49]

An extensive contribution to this proliferation of metaphor appears with Balfour Stewart and Norman Lockyer's 1868 two-part article "The Sun as a Type of the Material Universe." In their efforts to render the new energy physics broadly accessible, and indeed palatable, these authors give us an exemplary piece of Victorian popular thermodynamics. Economical in more ways than one, "The Sun" is a short but remarkably comprehensive sample of Victorian thermodynamic metaphor and its associated fantasies. Apparently written for the simple and disinterested purpose of elaborating the principles of the new physics and some of their consequences, the article is thoroughly laden with claims about the poetry and politics of science, as well as about the workings of class and the relations between nations. Even when Stewart and Lockyer seem most thoroughly engaged in recounting basic thermodynamic ideas and simple, familiar demonstrations, the language of economics is manifest. Thus, when they talk about the simple mechanical experiment of tossing a stone into the air, they note the transformation of kinetic into potential energy, the energy of motion into gravitational potential: "At the summit of its flight all the actual energy with which it started has been spent in raising it against the force of gravity." Similarly, we learn of the broader principle of conservation as we trace the transformations of the sun's energy into that of our fuel supply. "Nothing for nothing in these regions," they tell us, summing up the conservation of energy with impressive succinctness, but elaborating in specifically economic terms: "The sun's energy is spent in producing the wood or coal, and the energy of the wood or coal is spent (far from economically, it is to be regretted) in warming our houses and driving our engines."[50] The second law gives us causes to regret that are both physical and financial, because "far from economically" could as easily refer to heat loss due to friction as to the literal price of warming houses and driving engines. Their language, moreover, is typical. But their metaphor, extensive as it is, as well as their claims regarding the truth value of metaphor, provide not only a nice sampler of the broader, mutual production of thermodynamic and economic discourse, but also an indicator of how the logical tensions within the laws of thermodynamics themselves connect to a broader ambivalence about both economics and empire.

Small and Large

Why stop there? After all, the laws of thermodynamics obtain at every scale. From the interactions of molecules to the collisions between galaxies, we need only identify a closed system, and we find that energy is conserved while entropy increases. Thermodynamic metaphor seems early on to enjoy a similarly wide-ranging applicability. Seeking a unified theory of their own, Stewart and Lockyer move through local and individual economic concerns to claims about empire. Ultimately, their unification goals are even more lofty. They aim to develop, in a two-part article, a mechanism for intelligence and free will based on a hypothesis about the origin of sunspots and compatible with the principles of energy. To do so, they develop a principle of nonlinear effects that anticipates what contemporary chaos theory sometimes calls the "butterfly effect"—alluding to the notion that in nonlinear systems such as the weather, a small cause, say a butterfly flapping its wings in Tampa, may result in large and far-off effects, such as a tornado in Topeka. In the course of their argument, Stewart and Lockyer also seem to partake of what chaos theory calls "self-similarity," a principle that accounts for phenomena at vastly different scales looking very much alike. Why, after all, are mountains shaped like mole-hills?[51] For Stewart and Lockyer, this self-similarity begins life as metaphor or analogy. Thus the sun figures as "a type of the material universe." In accounting for the development of sun-spots, they hope to pave the way to explain nothing less than "the Place of Life" and the possibility of free will in "a Universe of Energy." Not only do the laws of energy obtain at solar as well as personal scales, so does the property they designate "delicacy of construction." From this property they conclude that "the bond between members of our system appears to be a more intimate one than has hitherto been imagined."[52]

As they insist on the truth value of metaphor/analogy, the distinction between metaphorical likeness and the (self-) similarity of systems at different scales—or in apparently different realms of inquiry—shrinks to the vanishing point. They observe "a striking likeness between principles which nevertheless belong to very different departments of knowledge." Though they can't seem to settle on what to call it, whether "a unity in . . . variety" or "a community of type," this likeness accounts for the similarity between systems at every scale. And they insist that this resemblance is not "a merely fanciful one, or one which the mind conjures up for its own amusement."

Before, however, we get to the likeness between sunspots and living beings, Stewart and Lockyer spend a considerable amount of time on a system at a scale in between them. What they call the social world refers initially to the relation between individuals of different classes but ultimately also includes that between nations: "We shall venture to begin this article by instituting an analogy between the social and the physical world." Though "instituting an analogy" suggests they intend to create a mental construct that serves human understanding rather than claiming they've discovered a likeness with an independent existence, their belief that "analogies . . . ought to be not fictions but truths" is reiterated in the language of intimacy. This statement refers, it seems, not only to the close bond between sun and planets, but also to those between social and physical spheres of inquiry: "We see from this how intimate is the analogy between the social and the physical worlds as regards energy." And apparently, the only thing that distinguishes this intimacy from the physical intimacy that characterizes the solar system is that "in the former it is impossible to measure energy with exactness." It depends simply on our capacity to measure. As analogies come to seem closer to truths, measurement becomes subject to doubt because of human limitations. Indeed, measurement may well turn out to be what "the mind conjures up for its own amusement."[53]

As they elaborate its likeness to the physical world, Stewart and Lockyer make quite a few claims about the social world. In doing so, they recall the roots of the language of energy in social contexts, whence it finds its way into the physical domain. Their social claims are thus presented not so much as claims, as what is already known, obvious, *factual*: "Energy in the social world," they assure us, "is well understood." As they introduce the term *energy*, it attaches to an individual we all know: "When a man pursues his course undaunted by obstacles he is said to be a very energetic man. By his energy, we mean the power he possesses of overcoming obstacles." An impressive fellow, no doubt, but also the way the authors connect energy to another important physical quantity called *work*: "The amount of his energy is measured by the amount of obstacles which he can overcome, by the amount of work he can do." Of course, there would be no drama without obstacles. Shortly, we find the energetic man's endeavors subject to run-ins with another man, who though he "may be deficient in personal energy, yet he may possess more than an equivalent in the high position which he occupies." In this way, we move through the familiar notions of personal energy and energy of position—phenomena presented here simply as facts—to the concepts we will eventually come to know as kinetic and potential energy: "Thus we have also in the physical world two kinds of energy: in the first place, we have that of actual motion, and in the next we have that of position."[54]

We also have some rather interesting suggestions regarding the classes. Apparently "turn[ing] to the physical world," the authors never leave the social behind. As they move to that classic mechanical example of the exchange between kinetic and potential energy—shooting a stone upward—we are reminded of their own casual suggestions of the violence inherent in the class struggle they depict. For the energetic man, it seems, is at war with the man of position. The energetic man himself "may in truth be regarded as a social cannon-ball. By means of his energy of character he will scatter the ranks of his opponents and demolish their ramparts." But he may well be defeated by his opponent, whose "position . . . enables him to combat successfully with a man of much greater personal energy than himself." It is then that classical physics and class struggle most thoroughly intersect: "If two men throw stones at one another, one of whom stands on the top of a house and the other at the bottom, the man at the top has evidently the advantage."[55] Evidently. This is Newton. This is the stuff we all know already. And it serves to reinforce the new concepts of energy as well as Stewart and Lockyer's particular spin on class relations. Though we find some more, some less well equipped for the struggle, sometimes with the accoutrements of modern war at their disposal, other times reduced to throwing stones, the fight is at all times subject to the laws of energetic exchange.

The quiet dig at the virility of the upper-class man, who though advantageously placed is sadly deficient in personal energy, leaves us momentarily in expectation of some more elaborate, even progressive, critique of the class system. But the picture that develops is more ambiguous than this. For the energetic man, a new position in the world may be a mere stone's throw away, but throwing stones is a decidedly difficult business, especially if one stands at the bottom. In this way, Stewart and Lockyer's picture of class mobility is laden with social and physical inertia. Today's energetic man may rise, though not without difficulty, because gravity functions as "that force which keeps a man down in the world." But in doing so he reiterates, even justifies, the current status of his opponent, who had a similarly energetic ancestor: "The founder of the family had doubtless greater energy than his fellow-men, and spent it in raising himself and his family into a position of advantage." This fact of class history derives from an energetic principle: "High position," as the authors ambiguously put it, "means energy under another form." Energy cannot be created or destroyed, so the current position of the family must derive from some previous expenditure of energy; "at some remote period a vast amount of personal energy was expended in raising the family into this high position." And even though "the personal element may have long since vanished from the family," the law of conservation tells us that "it has been transmuted into something else"—the energy of position. The man of position thus rests on the efforts of his ancestors and may "accomplish a great deal owing solely to the high position which he has acquired through the efforts of another."[56] It is significant, however, that the energetic ancestor rose through exactly the same mechanism and under the same conditions as today's energetic man would do. Stewart and Lockyer thus suggest the vulnerability and potential decay of the upper classes even as they rationalize their current status; they endow the current hierarchy with impermanence while they naturalize these developments as something governed by no less a natural law than that of gravity. The more things change, they seem to suggest, the more they stay the same. Such double-edged gestures, I would suggest, are typical of the dual claims of Victorian thermodynamic metaphor. Progression appears repeatedly against the background of conservatism. This characteristic doubleness inheres in the shape of thermodynamics itself—its foundation of two laws, its double building blocks of energy and entropy—and derives from its development along multiple fronts.

God and Politics, Conservation and Change

So it seems that for a short time in the middle of the nineteenth century—before it is popularized as the harbinger of doom—the discourse of energy manages to reconcile the idea of a world that changes with a conservative impulse to preserve and naturalize the status quo. We have already seen treatments of empire that have no beef with the system itself, though they may identify better and worse ways to run it, or even wish to shift the hands in which power (imperial, scientific, social) may reside. Similarly, we see in Stewart and Lockyer a depiction of class mobility that leaves the infrastructure firmly in place. This reconciliation of what seem to be antagonistic progressive and conservative gestures is built into the science of energy physics in the tension that forever inheres in the first and second laws. It is why these laws can provide consolation for the disturbing tendencies of Darwinism to suggest instability throughout the physical world and, by implication, causality, or analogy, throughout the social as well.[57] Due largely to its deployment by such self-proclaimed agnostics as T. H. Huxley, Darwinism further seems to support a worldview without (an actively involved) God. What we now call physics, of course, had long shown tendencies in this direction; Pierre-Simon Laplace is reputed to have said to Napoléon, when asked why God did not appear in his *Celestial Mechanics*, that "I have no need for that hypothesis." The science of energy, however, especially as articulated by certain proponents systematically divorcing themselves from scientific naturalists, takes another route.

A quick return to Joule and Thomson suggests how the thermodynamic tension between conservation and change is attached to their hypotheses regarding God. Joule grounds his principle of conservation in religious conviction (see chapter 1). As he elaborates, it becomes clear how his notions of conservation tie to a broader conviction that all's right with the world: "Thus it is that order is maintained in the universe—nothing is deranged, nothing ever lost, but the entire machinery, complicated as it is, works smoothly and harmoniously."[58] Joule's lifelong political conservatism, combined with his family's rise in social status (not unlike the energetic man's) and concomitant embrace of Anglicanism,[59] suggest a desire for conservation in the face of evident change. As I have argued, his belief in political and religious conservation must be painstakingly reconciled with the loss/change to which his experiments bear witness. Although "every thing may appear complicated and involved," and certainly his experiments suggest the "apparent confusion and intricacy of an almost endless variety of causes, effects, conversions, and arrangements," especially the very confusing "heat loss due to friction," he is nonetheless convinced that "the most perfect regularity [is] preserved—the whole being governed by the sovereign will of God."[60] That he sustains an idea of order, harmony, and even perfection, in the face of disorder leads to frequent interpretations of Joule as a conservative writer. His principle of conservation is fully implicated in these other forms of conservatism, all of them sustaining themselves alongside considerable experiential evidence of change.

Though fully on board with the idea that God is uniquely endowed with certain talents for creation and destruction, William Thomson also maintained religious ideas emphasizing change as part of God's order. Scottish Presbyterian notions of transfiguration and gift giving underlay his early articulations of what would become the second law. His reworking of

Carnot's theory of heat engines depended on "a fundamental conviction that 'Everything in the material world is progressive.' The directional flow of energy through space offered human beings the opportunity of directing, though not restoring, those mighty gifts of the Creator, the energies of nature."[61] Thomson's notion of "progression" would change (degrade?) to one of "dissipation" over time—a less upbeat alternative. But the formulation "everything in the material world" would linger in Thomson's conviction that something outside this world, or at least outside our understanding of it, "in the great storehouse of creation," would intervene to release us from the consequences of universal decay. But in both aspects of the notion, change was central. In what God would do, it was possible; in what we could do, it was inevitable.[62]

Stewart and Lockyer's essay suggests how such notions may be revamped and extended in the interests of popularization and promotion. As they expand their unification efforts, they extend the mechanism that accounts for free will among living beings. A similar mechanism, they argue, would allow for a "Supreme Intelligence [who] without interfering with the ordinary laws of matter, pervades the universe, exercising a directive energy capable of comparison with that which is exercised by a living being." Like Thomson and Joule, they wish to reserve space for a God consistent with the laws of thermodynamics, involved with the known universe, but uniquely situated. And in positing "a region beyond our ken in which energy [may be] created,"[63] they not only echo Thomson's "great storehouse of creation," but also anticipate Stewart's coauthorship of *The Unseen Universe*.

This concern for the divine by no means precludes more mundane concerns. Indeed, they ally themselves with Joule and Thomson in these as well. Their statement of "the grand principle of the conservation of energy" streamlines the history of its discovery, reducing it to "a principle lately proved by Dr. Joule."[64] This apparently simple gesture, as it happens, is a highly politicized act. Combined with their various efforts to reconcile religious belief with energy physics, the attribution to Joule reveals the authors' preference for the North British school over Cambridge efforts (especially Tyndall's) to connect energy physics to scientific naturalism and to give credit elsewhere. Their phrasing "a principle lately proved" glosses over the religious commitment to conservation that pervades this group of scientists of energy. Similarly, their choice of "Dr. Joule" over Thomson's preferred "Mr. Joule of Manchester" seems to wish to highlight academic status over Joule's social qualifications.

Like anyone who wants to be taken seriously in nineteenth-century English science (including the North British school), Stewart and Lockyer make efforts to show themselves followers of Newton. Beginning with the case of throwing a stone, they evoke the law of gravity in an uncomplicated segue to the laws of energy. They neatly posit the conservation of energy as resembling the more familiar conservation of matter, in spite of the fact that the conservation of matter is being pushed out of scientific significance. "Energy," they say, "like ordinary matter, is incapable of being either created or destroyed." And when they move to the second law of thermodynamics, they clearly wish to establish the laws of energy not only as successors to Newton's, but also as reflecting common sense, in firm opposition to the long-abandoned notion of perpetual motion, for "the law of the dissipation of energy shows us at once why a perpetual motion and an ever-burning light are both equally impossible."[65] They thus embrace Thomson's term *dissipation*, which already resonates with moral and political implications when Thomson takes it up. In fact, the "earliest English uses" of the term "refer to political contexts: 'sub[v]ersions of empires and kingdoms, scatterings and dissipacions of nations' [OED]."[66] Thomson himself adopts the term following an 1842 letter from his father that bemoans the doings of a certain Glasgow student recently found "to have been indulging in habits of dissipation."[67]

Like *dissipation*, the term *degradation* is closely associated with the second law, as well as with moral, physical, and social decay. Stewart and Lockyer use *degradation* in their first statement of the second law, where they find once more a "striking analogy between the social and the physical world." Their introduction of energy degradation is so bound up with their insistence on the validity of analogy and claims about the social world that it's hard to recognize as a physical claim: "For as in the social world there are forms of energy conducing to no useful result, so likewise in the physical world there are degraded forms of energy from which we can derive no benefit." A man, in fact, "may degrade his energy." These claims feel a bit different from Stewart and Lockyer's statement of "the grand principle of the conservation of energy," which tries, anyway, to emphasize its purely physical validity. Certainly, when they evoke such phrasing as "the law of dissipation of energy," they seem to highlight the physical world and scientific antecedents. But the connection they make between social degradation and the degradation of energy highlights the impossibility of separating the two ideas. Moreover, "in both worlds, when degradation is once accomplished, a complete recovery would appear to be impossible, unless energy of a superior form be communicated from without." Hints of the notion of superiority linger even in claims that seem to leave the social world behind. The law of dissipation asserts that "there is a tendency in the universe to change the superior kinds of energy into inferior or degraded kinds, which latter can only to a very small extent be changed back again into superior forms."[68] Thus energy itself seems to maintain its own social hierarchy even as we step away from the explicit concerns of social mobility and class struggle.

Tendencies Abroad

And of course, as so many have before, Stewart and Lockyer keep hoping that "energy of a superior form [will] be communicated from without." Where, however, is *without*? Undoubtedly reinforcing the notion that hierarchy in the social world reflects and reiterates the relation between the divine and the physical, several of Stewart and Lockyer's constructs suggest more mundane boundaries. Their statement that inferior energy is "degraded" by virtue of the fact that "we can derive no benefit" from it suggests that "we" and our concerns are the measure of the level of degradation of physical energy. It follows that "our existence and well-being depend on the presence in the universe of a large quantity of superior energy" and "it becomes us to take stock . . . of the goods that have been placed at our disposal."[69] Suddenly, we seem to have access to Thomson's "great storehouse." But this still begs the question of who *we* are. Unless we incorporate the whole of the universe, any question of how or whether there is a net degradation of energy depends critically on where we draw boundaries. What is within that may or may not receive energy from without? Who has the goods at their disposal?

There are those, it seems, working in ways detrimental to our existence and well-being: "There is, in fact, a tendency abroad to change all kinds of energy into low-temperature heat equally spread about,—a thing that is of no possible use to anyone."[70] Of course, Stewart and Lockyer here intend to restate the second law; energy tends to transform itself into diffuse, relatively slow, molecular motion, aka low-temperature heat. And their relatively passive construction—"a tendency . . . to change"—is consistent with the physical principle that energy degrades automatically into low-temperature heat (just as it conserves itself without the aid of a conserver). But this claim conspicuously avoids the reflexive, as in "the tendency of energy to change into" something or even simply "to change itself," and still seems to imply an external agent, though without actually identifying who that might be. Low-temperature heat "equally spread about" sounds almost communal or democratic, such that when they claim that "it is of no possible use to anyone," the physical and social implications of the claim split. Where *we*, in the broadest sense, all suffer from the degradation of energy, the referent of *we* seems, socially at least, somewhat less than universal. Their concern regarding "our existence and well-being"[71] then extends the move that is at once physically inclusive but socially divisive. We have, of course, seen this social splitting before in the battle between the man of great personal energy and the man who possesses energy due to his high position. Both are possessed of—and subject to the laws of—energy, but in different forms. And the difference that remains within the universal principles is thus thoroughly naturalized—as natural as (though not identical to) the difference between the superior and inferior; both are energy, but in different forms.

This moment, however, also introduces a slight shift in the economic metaphor that pervades the discourse on energy. "A tendency abroad to change" may imply a general social movement, or, alternatively, it could suggest the lamentable impulses of foreigners. And it is shortly hereafter that Stewart and Lockyer start to shift to concerns akin to Carnot's—concerns less about domestic class relations and more focused on the relations between nations. Of course, they do so in the interest of physical speculation, where their tendency is to reduce rather than expand the scale of their concerns. Finally, they have arrived at their avowed purpose: to propose an analogy between solar physics and the mechanics of living bodies. By applying the principle of "delicacy of construction" at this much smaller scale, they propose a condition consistent with the principles of energy physics under which living bodies may exhibit free will. Certainly living bodies cannot create or destroy energy, but if endowed with "infinite delicacy of construction" they may, with an amount of directive energy smaller than any measurable quantity, be able to produce tangible results. This infinitesimally small amount of directive energy, we take it, is the energetic equivalent of intelligence or free will, inaccessible, unpredictable, but physically conceivable.

To conceive it, they propose several machines that might present analogies to living bodies—machines in which very small triggers may yield very large results. These machines shift their political metaphors from the local to the international. Imagine, they suggest, "a gun loaded with powder and ball, and very delicately poised," a gun with a hair trigger. In such a case, "the expenditure of a very small amount of energy upon the trigger [will achieve] a stupendous mechanical result." This result is not only stupendous, but also highly suggestive: "If well pointed, it may explode a magazine,—nay, even win an empire." That's quite a claim for one gun, but it gets better. For they don't limit such effects to guns. They imagine a delicate electrical arrangement, in which an imperceptibly small amount of what they call "directive energy"—in fact, an amount "less than any assignable quantity"—will start an electrical current. They continue their physical speculations in a manner somewhat disconcerting to an unsuspecting American reader, such that the current, thus "made to start suddenly, [will] cross the Atlantic and . . . explode a magazine on the other side."[72] Thus with an amount of energy too small to measure, the power of the magazine is turned on its possessors. Compared with the energetic potential that inheres in poise and delicacy, the mere stockpiling of arms seems crude and shortsighted. And it is clear who is poised on the proper side of the Atlantic. Indeed, this subtle critique of American power recalls the critique implicit in *The Coming Race*, where the American narrator represents a place (both the United States and the upper world) blessed with greater sources of energy, but nonetheless threatened by a nation better versed in its more subtle uses. Thus the authors develop not only a physical loophole that allows for the imperceptible workings of free will, but also a barely perceptible fantasy of the continued growth and expansion of British power in spite of competing claims from across the ocean.

The progress of empire, even civilization, suggested by Stewart and Lockyer's hypothetical machine, replays that triggered by the development of a more familiar machine: the steam engine, of course. In keeping with their contention that analogies ought to be not fictions but truths, Stewart and Lockyer provide us with yet another instance of "the wonderful principle of delicacy which appears to pervade the universe of life":

We see how from an exceedingly small primordial impulse great and visible results are produced. In the mysterious brain chamber of the solitary student we conceive some obscure transmutation of energy.

Light is, however, thrown upon one of the laws of nature; the transcendent power of steam as a motive agent has, let us imagine, been grasped by the human mind. Presently the scene widens, and as we proceed, a solitary engine is seen to be performing, and in a laborious way converting heat into work; we proceed further and further until the prospect expands into a scene of glorious triumph, and the imperceptible streamlet of thought that rose so obscurely has swelled into a mighty river, on which all the projects of humanity are embarked.[73]

The workings of this single, highly delicate machine range in scale from the barely perceptible workings of the human brain through "all the projects of humanity." The first, delicately poised, serves as the trigger for all the rest. All of the workings of the machine, moreover, consist in the literal transformations of energy. When the solitary student "conceive[s] some obscure transformation of energy," he does so—according to the authors' favored model—through the mechanism of an obscure transformation of energy, which is conception itself. It is perhaps this transformation of energy in his brain chamber that "we conceive." Alternatively, perhaps we, our class, or our nation simply share in this not-so-solitary student's conception; his idea is our idea. A truly Enlightenment vision of the progress of science follows, as light is thrown (metaphorically, if we can still make such a distinction) on a law of nature. This light, though, is but a step in the chain of transformations in which energy manifests itself as thought, light, heat, and triumph.

This vision of progress echoes some we have seen before. Certainly the progress of thought as swelling streamlet recalls Stewart and Tait's "wave swelling as it advances shoreward." But here, this progress of the empire of thought is even less readily disentangled from the more tangible progress of empire. As the "prospect expands into a scene of glorious triumph," we recall the "expansive force" of steam in Carnot's formulation. This expansive capacity, itself a transformation of thought, strongly echoes Carnot's model of the progress of empire. There too, the development of the steam engine—triggered by specifically British thought—has similarly far-reaching results: through the avenue of mighty rivers, the world is transformed into one unified, civilized, technologized nation. Here, they serve to connect "all the projects of humanity"; there, "to unite the nations of the earth as inhabitants of one country."[74] As we move from the man who is the "social cannon-ball" to a shot heard round the world, the social metaphors escalate from individual and class struggle to international politics. Clear as a bell on the always/already social character of energetics, Stewart and Lockyer are engaged in the project of promoting these words for precise scientific uses. Nonetheless, the social metaphor, however subordinate to this larger goal, shapes a discourse on economic change and anxiety, justifies class distinction even as it suggests its impermanence, and finally explodes in international politics.

4   A Far Better Rest: Equilibrium and Entropy in
A Tale of Two Cities

An enlightened people, and an energetic public opinion . . .
will control and enchain the aristocratic spirit of the government.
—Thomas Jefferson[1]

With fire, everything changes.
—Michel Serres

Why Dickens? Why Now?

In 1858, Herbert Spencer wrote to John Tyndall, communicat-
ing his distress at the implications of the new energy physics
for his "doctrine of ultimate equilibrium."[2] Having built his
system of political philosophy on the premise that equilibrium
is the necessary and proper goal of social change, Spencer was
understandably dismayed to find that thermodynamically
speaking, equilibrium or *rest* is death. In 1859, Dickens published a different kind of narrative—less self-conscious, perhaps,
than Spencer's, but similarly dismayed about the implications
of thermodynamics for social stability. Boasting what may
be the best equilibrated opening in the history of the novel, *A
Tale of Two Cities* is nonetheless fraught with thermodynamic

instability. It is a narrative that wrestles with the relations between force, heat, energy, and work, a narrative that counts the entropic costs of attempting order and that fears the inevitable loss entailed in the exchange.

As Dickens spins his tale of a family swept up in the currents of the French Revolution, he participates in the impulse to unify theories that we considered at length in chapter 2. With *A Tale of Two Cities*, he gives us a narrative extending the laws of nature to the actions of individuals and societies, to characters, and to plot. As our look-alike heroes Charles Darnay and Sydney Carton make their way through a world driving, as ours is, to ever greater entropy, the struggles of the former are answered by the energetic innovations of the latter. The simultaneous forces of nature and culture embroil Darnay in the chaos of revolution, nearly bringing him to the guillotine. But (spoiler alert!) Carton engineers an exchange wherein he produces order at the expense of entropy elsewhere; he dies that others may live. In turn, Dickens's broad application of energy principles suggests that fiction, far from being the opposite of fact, is engaged in the kind of fact making that Bruno Latour traces in scientific articles, the kind we have already seen in poetry and popularizations. "Listeners," Latour tells us, "make sentences less of a fact if they take them back where they came from, to the mouths and hands of whoever made them, or more of a fact if they use them to reach another, more uncertain goal."[3] Where *The Coming Race* may be said to have sent certain scientific statements upstream, producing, for example, "the correlation of forces" as artifact by invoking Faraday himself and staging the controversy attached to "force," Dickens uses energy very differently. In *A Tale of Two Cities*, energy principles are "incorporated into tacit knowledge with no mark of . . . having been produced by anyone."[4] What could be more tacit than hardly mentioning them at all? Indeed, it takes a special lens even to see them at the base of Dickens's various statements about revolutions, families, and individuals. They are simply the way the world works at every level. And thus treated as facts, they become more factual.

I should probably stop here and qualify. I would not for the world wish to suggest that any amount of fictioneering will enable us to defy the second law or that saying so enough times will ever *get* energy to conserve itself. But facts are human things. They are statements we—collectively and after considerable trial—hold to be true. Talented though he may be, Dickens cannot single-handedly make a fact out of spontaneous combustion simply by exploding someone in *Bleak House*. (More on his efforts to do so in the next chapter.) This is because (1) fact making is a collective process and (2) scientific facts undergo very specific kinds of trials, not all of which can be conducted in words. Fiction can, however, participate in scientific fact making and can do so quite effectively, while hardly seeming to do so at all. And thermopoetics, broadly speaking, is the widespread word work that went into the making of the facts of energy physics.

Equilibrium and Entropy

It was the best of times, it was the worst of times, it was the age of wisdom, it was the age of foolishness, it was the epoch of belief, it was the epoch of incredulity, it was the season of Light, it was the season of Darkness. . . . There were a king with a large jaw and a queen with a plain face, on the throne of England; there were a king with a large jaw and a queen with a fair face, on the throne of France. In both countries it was clearer than crystal to the lords of the State preserves of loaves and fishes, that things in general were settled for ever.[5]

Not too many scientific papers open like that! As we begin, *A Tale of Two Cities* seems to effect a balance of universal opposites and international rivals. But this beautifully equilibrated opening soon gives way, through a series of exchanges, to a far more limited, arguably degraded order. The opening lines suggest that a fantasy of social stability is extant, the fantasy that "things in general were settled for ever." As savvy readers of *The Coming Race* and *First Principles*, we know, however, that the stability that comes from *settling* is at best a mixed blessing, attaching as it does to the permanently stable, highly equitable, but not especially desirable state of "ultimate equilibration," total entropy, heat death. Through the contrived naïveté of his narrator, Edward Bulwer-Lytton manages to suppress this implication, sustaining an apparent fantasy of equilibrium throughout much of *The Coming Race*. Similarly, Herbert Spencer, though fully aware of this complication, carefully segregates his discussion of its implications, preserving for huge portions of *First Principles* the impression of progress that equilibrium belies. In Dickens, on the other hand, the fantasy of equilibrium barely survives the satire of the opening lines. And almost immediately, any glimmerings of such a fantasy give way to the anxiety of decay. In spilt wine and in spilt blood, Dickens makes it clear that no fantasy of stability can survive the drive to disorder. Even in England "there was scarcely an amount of order . . . to justify much national boasting." Heat itself seems to be the culprit where the fantasy is undermined by the burning of people and of pamphlets. Elsewhere, as France "roll[s] with exceeding smoothness down hill, making paper money and spending it," while the horses of the Dover mail "[deny] that the coach could be got up the hill," the novel's energetics evolve, equally applicable to things economic as to the work that must be done against gravity.[6]

Finally, the equilibration of the opening is exchanged for the lesser equilibrium in Carton's famous final words: "It is a far, far better thing I do, than I have ever done; it is a far, far better rest I go to, than I have ever known."[7] A localized echo of the opening, Carton's words, or rather what his words would have been, have neither the scope nor the length nor the authority of the opening passage. All of these have been sacrificed to the novel as heat sink. And the balance apparent in his utterance not only suggests a degraded manifestation of the energy of the novel's opening, but also replays the transformation of usable energy into its degraded counterpart, of work into heat, of "I do" into I "rest."

But the "far better rest" to which Carton goes suggests that not all entropic transformations are equal. Even as he enacts the drive to entropy, Carton's words comfort, suggesting that some at least of that energy has been or will be transformed in a useful way. Local order—in the utterance itself and in the domestic happiness that Carton's sacrifice ensures—is purchased, though at a cost. The work Carton does, moreover, is but the consummate instance of the principles of energy that drive the novel as a whole. As Dickens taps into an emergent chemical thermodynamics, which may affect even the identity of matter, we follow the transformations of force into La Force, carriages into tumbrels, hosts of innocent people into prisoners bound for the guillotine. But in the transformations that, after years of imprisonment in the Bastille, recall Doctor Manette to life, turning suffering into power, we find that all is not entropy, that locally at least, order and useful energy may be restored, though not without cost. Finally, *rest* (a term resonant with physical as well as with spiritual implications) itself proves but a component of a larger thermodynamic system, larger than the visible universe, as large as the novel itself.

Spencer's Equilibria

In repeatedly staging the tendency of things to roll downhill, to burn, or to become disorderly, Dickens confronts the problem of universal decay. He wrestles with the central anxiety wrought by the new energy physics: the anxiety of dissolution, entropy, or heat death. In this way, *A Tale of Two Cities* seems to replay a familiar thermodynamic sensibility.[8] And the idea that energy physics implies universal degradation is indeed hard to escape.

Spencer, too, acknowledges this problem of heat death as he develops his otherwise progressive model of the evolution of all things. In this model, he defines evolution as the integration of matter and concomitant dissipation of motion—a process that accounts for the presence of everything from the solar system to the functioning of individuals and the production and sustenance of advanced societies, each of these an inevitable consequence of Spencer's most basic axiom, the persistence of force. The evolution of species is but a midlevel instance of a far more universal process.[9]

This progressive, universal evolution dominates Spencer's interest. His explicit treatment of its equally likely counterpart—dissolution—surfaces largely in the latter portions of his *First Principles*, beginning with the chapter called "Equilibration." Dissolution itself proves simply the "reversal of those changes undergone during . . . evolution,"[10] and it is equally the inevitable consequence of the persistence of force. Clearly preferring to pursue the progressive consequences of the persistence of force, Spencer relegates dissolution to the sidelines. Still, he cannot but recognize the visions of heat death, of the universe in general and the sun in particular, which come to dominate so many Victorian popular understandings of energy physics. It doesn't matter that our calculations may be in error, that we don't know how fast "the Sun is expending his reserve energy." The fact remains that "this reserve of energy *is* being expended, and must in time be exhausted." This reserve energy is precisely what the sun has acquired during its evolution, but that is now "being slowly rediffused in molecular motion of the ethereal medium." This state of things cannot last forever, and when "all the relative motions of its masses shall be transformed into molecular motion, and all the molecular motion dissipated"—when, as a physicist might say, all mechanical energy has been transformed into heat of uniform temperature—the game is up. It is this state of random, uniform heat that constitutes what is known as the heat death of the universe, and what in Spencer figures as the "state of complete integration and complete equilibration . . . towards which the changes now going on throughout the Solar System inevitably tend."[11] This complete and final equilibrium marks Spencer's acceptance of a "position enunciated [to him by John Tyndall] that equilibration was death." Spencer further confesses in his letter to Tyndall that "your assertion that when equilibrium was reached life must cease, staggered me."[12] It is the point at which no more transformation of energy, no more work, may be done.

To accommodate the disillusionment thus wrought in correspondence with John Tyndall, Spencer distinguishes two forms of equilibration. It is only the "final equilibrium" that constitutes complete rest and universal death. His original "ultimate" equilibrium is now qualified as an interim state of "moving equilibrium," wherein "energy ever in course of dissipation, is ever renewed from without." Moving equilibration thus depends on the presence of something external, something larger—an outside to the equilibrated system from which it can draw energy. It is this equilibrium that enables the production and sustenance of heterogeneous, orderly, evolved systems. But sadly, it is only a temporary state, which in fact drives us toward a final equilibrium: "the fact it chiefly concerns us to note is that this process must go on bringing things ever nearer to complete rest."[13] Complete rest occurs when the external is itself exhausted and can provide no more energy; alternatively, complete rest figures as the inevitable state of a system that has no outside, when all distinctions of inner and outer collapse, and we have only one big, undifferentiated agglomeration.

Forces of Nature

Similarly, in Dickens, useful, productive, progressive work can occur only when one can get outside the system and draw on energies from elsewhere, but such a possibility arises, in Dickens as in Spencer, only after wrestling with and distinguishing it from the dangers of ultimate dissolution. *A Tale of Two Cities* is manifestly more interested in social systems and their decay than in natural ones.[14] But the narrative nonetheless alludes to natural systems with surprising frequency. From the "great grindstone, Earth" to the "force of the ocean" and even the "ray of light" that "mere human knowledge can split . . . and analyse the manner of its composition," the physical world figures as the subject of physics and mathematics, even that which we can reproduce in the laboratory, as the "whirlpool of boiling waters" or a "train of powder . . . suddenly fired."[15] And that which is subject to physical laws, tends inevitably in the direction of disorder, decay, and rest. Thus "the fountain . . . melt[s] away, like the minutes that were falling from the spring of Time"; "the sea did what it liked, and what it liked was destruction"; our earth shines but feebly; and by "the risings of fire and risings of sea—the firm earth [is] shaken by the rushes of an angry ocean."[16] Whatever the result of social upheaval, it is clear that the stars will be extinguished and "Creation . . . soon wrung dry and squeezed out!"[17] The exposition of natural processes is, however, far from Dickens's central concerns. Indeed, the functioning of nature (out in the universe or back in the lab) figures almost as axiomatic, certainly as known or knowable. It is familiar enough to function as the vehicle through which we come to understand even the most dismaying social events. So, we understand revolution as a cauldron, and "as a whirlpool of boiling waters has a centre point, so, all this raging circled round Defarge's wine-shop," the "vortex" toward which "every human drop in the caldron had a tendency to be sucked."[18] The physical world nonetheless bears more than an analogical relation to these events. Simile slides into metaphor, metaphor into indistinction. In the "positivist extension of natural law to human behavior," social processes become natural processes.[19] We understand the storm that takes the Bastille, because we understand the workings of water—both part of the same system, both subject to the same laws.

Boiling water in particular marries two tropes for nature's disorderly energy—water and heat—which appear again and again as Dickens develops a dynamics of revolution. And the fluid that figures as simile above, soon gives way to the fluid that *is* revolution. Defarge himself is borne along by the crowd rushing the Bastille. Indeed, "so resistless was the force of the ocean bearing him on, that even to draw his breath or turn his head was as impracticable as if he had been struggling in the surf at the South Sea."[20] Gradually, even the "as if" that divides the terms of the comparison vanishes. The shapes and faces of the revolutionaries constitute a "sea of black and threatening waters," an "ocean of faces" productive of forces of its own. This "remorseless sea" is less familiar than the workings of regular water, which is deployed as *fact*, as we try to understand the more uncertain "destructive upheaving of wave against wave, whose depths were yet unfathomed and whose forces were yet unknown."[21]

A second, equally destructive force of nature is similarly at work as revolutionary processes unfold. Like water, heat or fire figures throughout *A Tale of Two Cities* as a destructive and unstoppable force of nature. Fire acts more or less as we expect it to: "The château burned; the nearest trees, laid hold of by the fire, scorched and shriveled." Things get even hotter, as "molten lead and iron boiled in the marble basin of the fountain; the water ran dry." Eventually, physical processes figure as similes for other physical processes: "The extinguisher tops of the towers vanished like ice before the heat. . . . Great rents and splits branched out in the solid walls, like crystallization."[22] These processes are, moreover, essentially the same because both represent the changes of state wrought by heat, changes always from more to less ordered, solid to liquid or even—as in the water running dry—liquid to gas. "Like" things prove not merely analogous but identical in their driving physical principles. And this kind of likeness extends to processes we distinguish—at least superficially—as social rather than physical. The literal fires that raze the château merge with the apparently figurative fires in the breast and faces of Saint Antoine—one breast, many faces, a synecdochical slippage that suggests that what applies to the town applies similarly to its inhabitants. For the arrival of Monsieur Defarge and the mender of roads "had lighted a kind of fire in the breast of Saint Antoine, fast spreading as they came along, which stirred and flickered in flames of faces at most doors and windows."[23]

Together, then, the dynamics of fluids and of heat are the dynamics of revolution: "In such risings of fire and risings of sea . . . three years of tempest were consumed."[24] It is no accident that water and fire should dominate this figuration—as with the boiling waters above. Though apparently opposing forces, they figure together as forms of energy. One is mechanical, the other, thermal; both are similarly capable of productive work (evolution, ordering) or of becoming the repository of energy beyond its usefulness, the most productive and destructive forces, equal and opposite as action and reaction. Water and heat, moreover, have just become the subject of the same science, as energy physics accounts for the behavior of both and even the transformation of the force wrought by one into the energy of the other. Or, as Michel Serres puts it, "With fire, everything changes, even water and wind."[25] That here these drive, apparently inexorably, toward greater disorder nods to the ease of second-law degradation, the drive to destructive or final equilibrium apparent in revolutionary processes. And entropy lives up to its name, which "borrow[ing] the root of the term 'trope'" acts as metaphor,[26] transforming like into like, in the interests of conservation and comprehension.

Rank Dissolution

Monseigneur—that figure of class oppression—proves a decidedly eighteenth-century thinker, oblivious to the broad applicability of thermodynamic principles. The irony with which the narrative treats him, however, stages the dismay wrought by the new physics, a dismay that in Monseigneur's case is simultaneous with that wrought by economic and political upheaval. Acknowledging, however casually, the Victorian anxiety that the stars will go out, Monseigneur nonetheless sets himself above the laws that govern all the rest of the universe—not only moral laws, but also physical laws, which (as Dickens suggests and Spencer insists) are at the base of all others. Not surprisingly, the great and very personal fact of dissolution comes as a bit of a shock. It seems "strange that Creation, designed expressly for Monseigneur, should be so soon wrung dry and squeezed out!" Monseigneur is thus taken by surprise by "a phenomenon so low and unaccountable,"[27] because he has been clinging to the fantasy of perpetual motion—something that Newtonian physics doesn't disallow, but that the second law of thermodynamics makes impossible. For in the year 1780, "who among the company at Monseigneur's reception . . . could possibly doubt, that a system rooted in a frizzled hangman, powdered, gold-laced, pumped, and white-silk stockinged, would see the very stars out!"[28]

In this way, Monseigneur represents not only a moment in the history of nations, but also one in the history of science. Though classist and emphatically outmoded, there is nothing anti-Newtonian in his fantasy of perpetual motion, in his belief that "such frizzling and powdering . . . would surely keep anything going, for ever and ever."[29] There is, however, something fundamentally unscientific in his denial of cause and effect—his willingness "to talk of this terrible Revolution as if it were the only harvest ever known under the skies that had not been sown."[30] The novel, moreover, ties this antiscientific stance to his denial of universal heat death. Monseigneur's "extravagant plots . . . for the restoration of a state of things that had utterly exhausted itself, and worn out Heaven and earth as well as itself" is not only politically, but physically misguided as well, ignoring as it does what Maxwell calls "our experience of irreversible processes."[31] And if experience were insufficient, the implication of Monseigneur's belief that useful energy can come from a system that is "utterly exhausted," can come as it were from nowhere, patently defies the second law.

Unlike Monseigneur, Spencer would have seen it coming. He would agree, however, that "social disorders or disasters . . . so far from constituting further Evolution, are steps toward Dissolution."[32] Spencer, moreover, shares some of Monseigneur's dismay at this decay—though not his surprise. Under ordinary circumstances, "citizens are bound up into distinct classes and sub-classes." Disaffection results in political outbreak; agitation eventually "fuses ranks that are usually separated. Acts of insubordination . . . tend to obliterate the lines between those in authority and those beneath them." The kind of equilibration suggested by the leveling of rank and the elimination of social distinction can only be dissolution, tending toward Spencer's final equilibrium—rest, disorder, death. The loss of distinction, moreover, occurs within as well as between classes: "Arrest of trade causes artisans and others to lose their occupations; and ceasing to be functionally distinguished, they merge into an indefinite mass." Spencer's anxiety of indistinction derives from a thermodynamic sensibility that Monseigneur is wholly lacking.

It is a consequence of the assumption—implicit in Spencer's first principles—that society functions like an engine; difference is required to do work. (Dickens, as we will see in the next chapter, pursues this premise at great length.) The collapse of society is energetically construed as the loss of this necessary distinction. As artisans and others cease to be "functionally distinguished," we are reminded that function requires distinction. The fusion of rank indicates further the loss of distinction or differentiation that enables work. The failure of power and the arrest of trade inevitably follow, as by a law of nature: "When at last there comes positive insurrection, all magisterial and official powers, all class distinctions, all industrial differences, cease: organized society lapses into an unorganized aggregate of social units."[33] As with famine and pestilence, rank dissolution is decidedly entropic, a change from order to disorder.

Spencer thus inclines toward a well-articulated hierarchy, at least in part because he believes it is the only way to accomplish work. By contrast, the distinction of rank preferred by Monseigneur never was conducive to work. The state of France at the opening of *A Tale* thus suggests this energetic qualifier: that difference is merely a necessary and not a sufficient condition for producing work. Not every hot spot implies an engine. Where social differentiation reveals itself as "glaring . . . disparity"[34] rather than functional distinction, it may do no more than drive rapidly toward entropy. Against such a drive, Monseigneur can do no more than Madame Defarge, who proclaims, "Tell the Wind and the Fire where to stop; not me!"[35] Monseigneur has achieved neither perpetual motion nor a moving equilibrium driven from without, but merely fixity. And, as Spencer predicts, "Decay follows a fixity which admits of no adaptation to new conditions."[36] Dissolution hits Monseigneur hard; "exquisite gentlemen...were in the most exemplary state of exhaustion."[37] Both the class and the individual are subject to the same decay that drives the universe; "the sun and the Marquis [go] down together."[38]

Darnay Bound

The dynamics of revolution and the forces of nature see to it similarly that the son, or at least the nephew, and the Marquis will go down together. Darnay is a particularly disturbing figure to Monseigneur in general, and to his uncle the Marquis in particular, because he makes it clear that the energies that have sustained the illusion of perpetual motion have all but run out. He is the figure of inevitable degradation, the "degenerate successor," as one aristocratic refugee calls him, "of the polished Marquis who was murdered."[39]

Not only does Darnay thus evoke the acknowledgment that decay is extant, he also experiences the drive to dissolution firsthand. Each of his returns to France enacts the entropic transformation that is, by one definition, precisely the move from *free* to *bound*: "In an entropic process what is at first 'free' energy ('free' in the sense of available, ready to do work) becomes 'bound' energy (energy that, like the enormous amount of heat energy contained in the world's oceans, cannot be used to accomplish work)."[40] Not only in Darnay's move from freedom in England to imprisonment in La Force, but also in the earlier visit to his uncle, he represents the energetic as well as the political distinction implied by the free/bound opposition. In both senses, freedom entails the ability to do work. Darnay leaves France specifically to free his energies, "to live by his own industry in England, rather than on the industry of the overladen people of France." He finds himself, however, "bound to a system that is frightful to [him], responsible for it, but powerless in it."[41] He thus suggests that what is at stake is the relation of an individual to a larger system (like the world's oceans), which may function as sink rather than source of energy. Darnay seems to divine that France, as a system, may have no energy to spare and indeed may do no more than absorb his own.

This does seem to be the case. Within this larger system, a variety of forces work on Darnay, all contributing to the loss of what energies he has. Suggestive of energy in so many manifestations, "the winds and streams had driven him within the [magnetic] influence of the Loadstone Rock," closer to France and to his final rest. "The unseen force was drawing him fast to itself, now, and all the tides and winds were setting straight and strong towards it."[42] What was in England the potential energy that inheres in attraction at a distance, converts to motion as Darnay is drawn to France, and threatens to dissipate in disorder and destruction, Darnay's own bodily disintegration and the concomitant loss of his usable energies. Thus driven to his own destruction by the system to which he is bound, he becomes the figure of inevitable decay that the Marquis has long thought him. As such, he also forces the recognition that perpetuity requires a perpetuator. For Darnay's declaration of his absence of power prompts his uncle's vow to be that perpetuator, regardless of the draining of the individual within the system: "My friend, I will die, perpetuating the system under which I have lived."[43] But the persistence of force, as we will see, requires no such perpetrator, and in this, as in all things, leads inevitably to dissolution.

The Persistence of Force

The "persistence of force" resonates nicely with both physical and social implications—even as the distinction between these implications diminishes. As we have seen, Dickens repeatedly evokes the term *force* in his naturalization of revolutionary movement. What is at issue is a force—or the forces—of nature. *Force*, moreover, is a term that dominates physics from the time of Newton, and remains for much of the nineteenth century a contender for the quantity that will eventually become *energy*. It is also Spencer's favored term. His "persistence of force"—like the conservation of energy—depends critically on the capacity of force or energy to change forms. The recognition of this protean quality is at once what makes energy physics possible and what made it for so long elusive. That what we now call energy could take so many forms—including heat, light, electricity, gravitational potential, and mechanical effect or work—led for a long time to the impression that it could be created or destroyed.

As we saw in chapter 2, Spencer extends the claim of the new energy physics that "each force is transformable, directly or indirectly, into the others"[44] to include all of the forces of the known universe. Experiment has shown that gravitational attraction may transform into "sensible" motion, which may transform into "that mode of force which we distinguish as Heat [which] is now regarded as molecular ['insensible'] motion," and that any of these may become or derive from electricity, magnetism, light, or chemical action. Spencer figures that "throughout the Cosmos this truth must invariably hold." The "transformations of forces are . . . everywhere in progress, from the movements of stars to the currents of commodities." And from the other end of the spectrum, "if the general law of transformation and equivalence holds of the forces we class as vital and mental, it must hold also of those which we class as social."[45] From his most basic premise, the persistence of force, Spencer seeks to derive the laws that govern all natural and (what is hardly another category) human systems. In his many books, he tackles biology, psychology, sociology, and morality. In *First Principles*, he gestures larger and farther by briefly extending his argument at each stage, as we have seen, from the movements of stars, to those of commodities, to the social, the vital, and the mental. Spencer's usual pattern is to trace such similarities through systems in descending order of size (from the sidereal to the geological to the vital) and ascending order of evolutionary chronology (from the inorganic to the organic to the social). When it comes to his discussion of dissolution, however, he deviates from his usual pattern to treat the social first.[46] Thus for Spencer as for Dickens, social systems serve as the primary example of dissolution.

And of course, the principle that force persists is a central concern of *A Tale of Two Cities*. It is, moreover, a principle that depends critically on the capacity of force to transform. Dickens too traces the transformation of force and the dissolution that follows inevitably from it. Itself evidence of the persistence of force, *La Force* is clearly an echo of the Bastille, a new manifestation of the force once exerted under Monseigneur. Similarly, "Barsad, and Cly, Defarge, The Vengeance, the Juryman, the Judge, [are] long ranks of the new oppressors who have risen on the destruction of the old."[47] It is a third prison, however, that provides us with Dickens's best evidence of the persistence of force. It is an instance, moreover, that makes clear that all transformations resulting from the persistence of force entail degradation. In the tumbrels that carry prisoners from the Conciergerie to the guillotine, we see this economy of transformation clearly manifest. If Time could "change these back again to what they were, [they would] be seen to be the carriages of absolute monarchs, the equipages of feudal nobles, the toilettes of flaring Jezebels." This is, of course, not possible. Like the new physics, Dickens reminds us that Time "never reverses his transformations."[48] (Thus entropy earns its nickname of "Time's Arrow."[49]) Somewhat paradoxically identifying these emphatically changed tumbrels as "changeless" (as well as "hopeless"), Dickens emphasizes the direction rather than the absence of change. Evidence of both the first and second laws of thermodynamics, the changelessness of the tumbrels points rather to conservation in the face of transformation: what were carriages are now tumbrels, the same thing only different, transformed and degraded as (in spite of Dickens's sympathy with the downtrodden) the revolutionary Defarges mirror, replace, and are even worse than Monseigneur. Spencer observes similarly that "a political reaction never brings round just the old form of things."[50] Although in Spencer, this fact may result from the complexity of social transformations, in Dickens "never reverses" points to the apparent hopelessness of the second law. If these tumbrels are now changeless indeed, it is because they have reached their lowest, most degraded, most useless state.

The Mechanics of the Individual

As Dickens applies the same principles of entropic decay to the tumbrels themselves, as well as to the larger system of which they are part, we once again see the workings of metonymy and the unifying principles of energy. And in spite of Monseigneur's ignorance and Madame Defarge's emphatic denial, these principles apply to systems we class as individuals as well as those we class as natural or social. Somewhere between the movements of stars and the movements of molecules, we find the same principles at work in the mechanics that govern the movements of the individual. The "great grindstone, Earth" and the "lesser grindstone" on which the citizens sharpen their weapons,[51] find themselves echoed at yet another scale in the man who observes them. Jasper Lorry, who claims to "pass [his] whole life . . . in turning an immense pecuniary Mangle," is explicit about his own machinelike qualities: "Don't heed me any more than if I was a speaking machine—truly, I am not much else," he tells Miss Manette. "In short, I have no feelings; I am a mere machine."[52]

Looking "very orderly and methodical," Mr. Lorry is perhaps more self-consciously but not otherwise more mechanized than most. And his loud watch, in particular, ticks "a sonorous sermon . . . as though it pitted its gravity and longevity against the levity and evanescence of the brisk fire."[53] It thus replays the tension between Newtonian and Victorian physics. Its "gravity" suggesting both seriousness and Newton, this consummately eighteenth-century machine also evokes Paley's watch, and its implication of a divine watchmaker. Pitting its longevity, moreover, against the evanescence of the fire, the watch evokes the distinction between the possibility of perpetual motion and the surefire decay promised by the physics of heat. Thus not only is the watch itself a small machine, the watch-fire combination stages at a small scale the dismay that Spencer shares with Victorians in general that even the stars will go out, as well as Monseigneur's disappointment that he will not be around to see it. In spite of the gravity and longevity of Newtonian mechanics, the arrogance of Monseigneur, and the more benign confidence of Mr. Lorry's watch, the fire will out.

And like the watch and the larger machines of the universe and the political system, individuals prove subject to second-law degradation. Sooner or later, we will all burn out, like the lamps so readily replaced in the revolutionary fervor by burning officials such as Monsieur Gabelle, who observes "an ill-omened lamp slung across the road before his posting-house gate, which the village showed a lively inclination to displace in his favour." This exchange is not exclusive, but everywhere potential and once begun, inevitable and mathematically irrefutable: "whosoever hung, fire burned. The altitude of the gallows that would turn to water and quench it, no functionary, by any stretch of mathematics, was able to calculate successfully."[54] The impossibility of transforming gallows into water nonetheless highlights that another transformation—one in the direction of greater entropy—is taking place in the transformation of functionaries into lamps, bodies into fire.

Life itself consists both intuitively and thermodynamically in resisting this tendency to burn out. As Spencer understands it, "all vital actions" work together to balance a body's "internal forces" with those "external forces" that tend to bring "living bodies . . . into that stable equilibrium shown by inorganic bodies"—that is, death. The constant change that is "life, may be regarded as incidental to the maintenance of the antagonism."[55] Life, then, consists in resisting the state of final equilibrium, entropy or rest. Dr. Manette evinces an understanding of vital systems rather like Spencer's, as he diagnoses his own case: "There has been a violent stress in one direction, and it needs a counterweight."[56] Substituting this new balance of energies—what Spencer would call a moving equilibrium—for Newton's old action/reaction, Dr. Manette also seems to understand life, at least under strained circumstances, as the maintenance of an antagonism.

It follows almost intuitively that large-scale political reaction derives from too thoroughly experiencing life as constant antagonism. Both the external forces that drive us to our deaths and the internal forces that resist this tendency are manifest among the starving people of Saint Antoine: "All the women knitted. They knitted worthless things; but, the mechanical work was a mechanical substitute for eating and drinking."[57] Eating and drinking, as Spencer suggests, enable the transformation of energy from without into the integration of matter (growth) and mechanical work of the individual. In Dickens, however, we see that the usual vital actions of eating and drinking have themselves undergone a further transformation. As a manifestation of energy, work takes numerous forms. As the novel progresses, it becomes clear that knitting is a degraded form of work not only in the worthlessness of the product, but also in the larger energetic cost. Conducive perhaps to the local order of knots and stitches, this knitting is ultimately at the expense of greater disorder in the surrounding system. For Madame Defarge's knitted register of revenge ensures the death of so many whose names are recorded there; the guillotine effects a systematic disintegration of bodies before an audience of apparently perpetually knitting women, drawing their energy from a decaying system that can ill spare it.

In his periods of decline, Dr. Manette similarly substitutes one form of work for another: the mechanical work of making shoes substitutes for his former intellectual work. As he reflects on his attachment to his little cobbler's bench, he speculates that "it relieved his pain so much, by substituting the perplexity of the fingers for the perplexity of the brain, and by substituting, as he became more practised, the ingenuity of the hands, for the ingenuity of the mental torture." The energies of the body, it seems, must manifest themselves one way or another, and when one manifestation puts too much strain on the system, another route of less resistance is found. This transformation from intellectual to mechanical work repeatedly signals Dr. Manette's degradation, a regression to the decayed individual first recovered from the Bastille. On his failure to preserve Charles Darnay against a second set of revolutionary accusations, he again seeks the relief of mechanical work: "Let me get to work. Give me my work." Mr. Lorry immediately interprets this as decay, a return to "the exact figure that Defarge had had in keeping."[58] Strangely, however, Mr. Lorry's interpretation of a transformation in the direction of greater entropy or loss, seems also a backward movement in time. While the women's knitting never seems, even remotely, to lead to the women's eventually eating again, Dr. Manette's transformations seem reversible. Already "restored to life," he proves, even in his decay, more than usually capable of reversing the energetically ordained order of things.

Manette Transformed

The initial trend, of course, was downward. On first recovering her father, Lucie Manette traces the degradation of his energy, encouraging him to weep for his "useful life laid waste." But when she anticipates that they "go to England to be at peace and at rest," she presumes no particular restoration to usefulness. Indeed, like *waste*, the term *rest* signals (as in Spencer, above) the final equilibrium at which point no more work can

be done. Mr. Lorry, however—perhaps because of a lingering
Newtonian sensibility—retains hope for reversibility. In spite of
the "shadows of the night" and "the cold and restless interval,
until dawn," which may suggest entropy to the sensitized
Victorian, Mr. Lorry wonders not only "what subtle powers
were for ever lost" to Dr. Manette, but also "what were capable
of restoration."[59]

As suggested above by Spencer's "moving equilibrium" of
which "vital actions" are a subset, the reversibility that Jarvis
Lorry imagines is possible even in a worldview shaped by energy
physics, but with certain important caveats. Indeed, Dr. Manette
bucks the entropic trend, manifesting the transformation of
energy once lost or, more accurately, apparently wasted, back
to useful, ordered, purposeful energy: "The energy which had
at once supported him under his old sufferings and aggravated
their sharpness, had been gradually restored to him." In fact,
Dr. Manette becomes "a very energetic man indeed," meriting
the descriptor more than any other figure in the novel. The
energies that characterize him later in life, his "great firmness
of purpose, strength of resolution, and vigour of action," figure
as transformed or recovered manifestations of his earlier energies. And if "in his recovered energy he was sometimes a little
fitful and sudden," he nonetheless appears as evidence of the
conservation of energy that seems to have vanished and as a
test case for the possibility of raising the quality—the usefulness,
effectiveness, and power—of that energy.[60]

In this way, Dr. Manette proves the exception to the expected
rules that govern such transformations. The oddity, so disconcerting to Jerry Cruncher, of Dr. Manette's being "recalled to
life," anticipates the positive turn to the transformations of
energy he undergoes. In these transformations, Dickens empha-

sizes that this new energy, though differently manifested, is the same as what went before, "now . . . his . . . suffering was strength and power." In the "sharp fire" of his own ordeals, "he had slowly forged the iron" that could deliver his son-in-law, Charles Darnay. "It all tended to a good end," he assures his friend, "it was not mere waste and ruin." Jarvis Lorry registers the change in his friend's "kindled eyes, the resolute face, the calm strong look and bearing" as a restoration of energy as well. To him, Dr. Manette's life seemed "to have been stopped, like a clock, for so many years, and then set going again with an energy which had lain dormant during the cessation of its usefulness."[61] As a figure at once sympathetically and unimpeachably human as well as fully mechanized, Dr. Manette suggests the universal applicability of the mechanics of personal energy—the work that may be done toward positive transformation and the ease of negative transformation. In this way, he is an encouraging figure for human energetics. Unused energy proves not unusable—but only dormant. The energy that characterized the young Dr. Manette, we find in retrospect, had never been lost at all, though its usefulness had momentarily ceased. It now reappears, manifesting itself as "strength and power"—themselves, a new manifestation of what was once "suffering." "My old pain," he says, "has given me a power."[62] What appeared to be "mere waste and ruin" was tending somewhere after all.

Personal energies are thus posited as subject to the process of restoration, but that restoration, as we have seen before, seems to require an external agent—the aid of some external force, the winding that sets a clock going again or the act of kindling implicit in Dr. Manette's "kindled eyes." This external agent that once appeared to Dr. Manette in the figure of his daughter Lucie, now becomes Dr. Manette as he seeks to restore Charles to her. The movement of energy, implied by Dr. Manette's observation that "the relative positions of himself and Lucie were reversed," seems to suggest a simple form of conservation, as if energy simply passes back and forth between the elements of a closed system. But the system is not closed, and some energy loss can be traced. Thus the "aid of Heaven" suggests something external helping to drive the Lucie/Doctor exchange, and the subtle distinction between Lucie's "restoring me [Dr. Manette] to myself" and Dr. Manette's restoring only "the dearest *part* of herself to her" suggests that something less than all of the previous energy has been retained in a usable form. Thus, in spite of the apparent equivalence of the exchange, this is no perpetual motion machine. While certainly the restoration of Dr. Manette's energy disallows the feared possibility of "*mere* waste and ruin," it still suggests *some* waste and ruin.[63] Even with the aid from without, somewhat less useful energy remains than once was present.

Carton's Energy

Sydney Carton is, of course, no Dr. Manette. He seems to have no special dispensation within the laws of physics. Indeed, Carton seems far more emphatic a figure of decay than his double. Where Darnay does manage while in England to free his energies, to produce useful work out of his personal potential, Carton's energies seem largely useless. His friend Stryver first calls our attention to the uselessness of Carton's energy. A decidedly Victorian machine who "shouldered his way through the law, like some great engine forcing itself through turbid water," Stryver is explicit about Carton's energetic deficiencies. "Your way is, and always was, a lame way," he tells Carton. "You summon no energy and purpose."[64] Stryver, it seems, can summon what Carton cannot. The failure to use energy, however, by no means implies its absence. Certainly, we can see this distinction in the irony implied by Stryver's comment. After all, Stryver has managed to summon considerable energy and purpose from Carton; he quite profitably puts Carton to work. Nonetheless, Stryver's assessment of Carton's purposeless is reinforced by the narrative. Over and over, Carton is presented as a figure of degraded energy. "Waste forces within him, and a desert all around," Carton contemplates the might-have-been. But his hopefulness is fleeting, and he makes his way home, to wet his pillow "with wasted tears." For all that his energies conduce to nothing, however, the text is clear that they are not absent. There are "forces" within him. He is a "man of good ambitions and good emotions." He is merely "incapable of their directed exercise." Even his "wasted tears"—especially within the "desert all around"—become an image of energy that under other circumstances (another system? a more efficient machine?) could have been useful. An echo of Tennyson's "waste places," the whole of the city becomes the repository of energy past its usefulness, a heat sink, a desert, a wilderness, a "well of houses." Even the sun, the most prolific source of usable energy around, despairs in the face of this vast heat sink. "Sadly, sadly, the sun rose; it rose upon no sadder sight" than this man who antici-pates his own decay, who is "sensible of the blight on him, and resigning himself to let it eat him away."[65]

As we understand Carton's energy in terms of energy in other forms, we find a similar pattern. Shifting to energy manifest as light, we find that "the cloud of caring for nothing, which overshadowed him with such a fatal darkness, was very rarely pierced by the light within him"; again, there is light within, but it cannot pierce the darkness. And when Carton muses on the fluid dynamics we have seen elsewhere within the novel's energetics, he becomes quite self-conscious about his own energetic profile. Even as the sun's rising seems to belie his impression that "Creation [had been] delivered over to Death's dominion," Carton reiterates the uselessness of his own energies, as he lingers by the side of the Seine, "watching an eddy that turned and turned purposeless, until the stream absorbed it, and carried it on to the sea. 'Like me!' "[66] Like the desert all around and the cloud of darkness that surrounds him, the sea figures as the sink in which Carton's energies are absorbed rather than used; his forces become "waste."

But even Carton's recognition of his own waste energy is met by the glimmer of potentially useful transformation, as his subsequent prayers culminate in "I am the resurrection and the life"[67]—words that anticipate not only his death and subsequent burial, but also the potential usefulness of the transformation that is death. He too may be "recalled to life" in ways at once spiritual and energetic. Indeed, as we have seen, this confluence of religion and science is characteristic of much early energy physics, the belief that only God can create or destroy driving much of the development of the first law. In turn, as in *The Unseen Universe*, the conservation of energy provides for some a semiphysical avenue to the afterlife. Carton figures nicely within this larger physicospiritual narrative, because he becomes the site of a qualified alternative to the unrelieved drive to entropy, a loophole in the laws of energy physics. If, as a figure of decay, Carton reminds us that at the bottom of the entropy metaphor is the fear that "we are embattled residents of a universe that marches towards heat death," the far better things he does also suggest that we can "accomplish . . . a temporary and local reversal of a natural disorder that awaits us,"[68] that we can do work.

Carton's Work

From the start, we may see that Carton is a figure not only of
degradation, but also of transformation or exchange. In his first
appearance at court, he reveals the potential for exchange with
Darnay, which is fully realized in the novel's final moments:
"Allowing for [Carton's] appearance being careless and slovenly
if not debauched, they were sufficiently like each other to
surprise."[69] Carton's distinguishing disorder reminds us that
exchange is closely tied to degradation. And following this first
meeting, Carton actually identifies himself as a degraded manifestation of Darnay. Even as he decides there isn't much to like
about himself, Carton wonders why he should like a man who
looks like him: "A good reason for taking to a man, that he
shows you what you have fallen away from, and what you
might have been!" Suggesting that his own state is the result of
some process—"What a change you have made in yourself!"—or
the result of falling away from some better state, Carton muses
on but ultimately rejects the possibility that he can reverse the
process. Even if he could change places with Darnay, he doubts
that the effects would be similar, that Lucie Manette would feel
for him as she does for Darnay: "Change places with him, and
would you have been looked at by those blue eyes as he was,
and commiserated by that agitated face as he was?"[70]

But in spite of his doubts, Carton becomes a figure of positive
transformation. On the single occasion when he speaks openly
to Lucie, we see inklings of such a transformation; he appears
"so unlike what he had ever shown himself to be." This change
is decidedly counterentropic, as his hitherto useless energy
becomes directed and orderly: "From being irresolute and purposeless, [Carton's] feet became animated by an intention, and,
in the working out of that intention, they took him to the

Doctor's door." As Carton accounts for the possibility of such a counterentropic transformation, however localized, he acknowledges that he has drawn on an external source of energy. Like Dr. Manette, Carton finds this source in Lucie, to whom he confesses, "I . . . wish you to know with what a sudden mastery you kindled me, heap of ashes that I am, into fire—a fire, however, inseparable in its nature from myself, quickening nothing, lighting nothing, doing no service, idly burning away." Lucie's ability to kindle revives his energy into something that could be useful, at least while it burns. Carton acknowledges the relative usefulness of fire compared to ashes—that fire can quicken, light, do service, or to speak thermodynamically, can be transformed into work. At the same time, however, he insists that his fire will go the way of the "waste forces within" him and that he himself "draw[s] fast to an end." Miss Manette finds her own "power for good" insufficient to drive the counterentropic transformation.[71]

After all, Carton does draw fast to an end. But he has figured out how to do work en route. The usefulness of energy is relative; as water falling or fire burning may do work as it approaches a state of rest, so may Carton as he is driven toward death. Thus the process of decay—the move from more to less orderly energy, or from what physicists call a higher energy state to a lower— may itself do useful work. Orderly energy—a well-defined hot spot and cold spot—will become a disorderly lukewarm spot no matter what we do. Whether in the interim it drives an engine depends on us. As we will see in the next chapter, this possibility provides much of the hope—even the space for free will—associated with later Victorian energy physics. It is at the base of a more optimistic thermodynamic conception of life as well. Rather than something like Spencer's definition of life as the constant antagonism of forces, life may be understood as "a controlled burning, a pattern of energy flow."[72] As we descend, inevitably, toward a state of rest, life may well be that which best uses the fall.

Carton's insight is that he too can control the burn. The energy expended in his fall can be summoned and purposeful. As Carton himself informs Barsad, "Gradually, what I had done at random, seemed to shape itself into a purpose." As with Dr. Manette's, whose suffering is transformed into strength and power, we find Carton's energies transformed from a formerly useless state into a currently useful one, his "negligent reckless-ness of manner [coming] powerfully in aid of his quickness and skill."[73] In this way, Carton becomes, in his final encounter with Darnay, the external force that drives: "With wonderful quick-ness, and with a strength both of will and action, that appeared quite supernatural, [Carton] forced all these changes upon [Darnay]."[74] Thus, Darnay will profit from Carton as energy source. Carton's rescue, his sacrificing himself to the guillotine in Darnay's place, moves Darnay further from a final state of rest by infusing energy from without. Carton, for his part, is able to drive this exchange, the partial transformation of each man into the other, because he has harnessed to useful ends the energy expended in his own decline. Thus Carton's useful work, the thing he does, is simultaneous with the rest he goes to, far better because the energy expended in his fall is not wasted.

In this way, Carton becomes a small but consolatory loophole in the law of entropy. If all tends toward decay, perhaps we can make something of the journey. Against the background of the "crashing engine that constantly whirrs up and falls," Carton reveals that beneath the moving equilibrium of the guillotine, the apparently perpetual motion of its rise and fall, lies a more complex picture. Beyond it, of course, lies death, a final equilibrium, that great heat sink where he assures his companion, "there is no Time," where all transformation is at an end, and entropy, time's arrow, has reached its final and maximum state. On another scale, the guillotine will drive more and more disorder, oppression, and death, as "long ranks of new oppressors who have risen on the destruction of the old, [also perish] by this retributive instrument, before it shall cease out of its present use." Between this and universal heat death, however, lies another possibility. Carton's judicious use of his remaining energies drives the increase of order in a system larger than himself, as he restores the order of the family he cares about. At the same time, he offers hope that a similar counterentropic transformation may be in store for Paris as well: "I see a beautiful city and a brilliant people rising from this abyss."[75]

Spencer takes the trend one further, reading counterentropic transformations on the small scale (merely galactic) as indicators that even the final equilibrium may not be final. In spite of his own cautions against speculating on the unknowable, Spencer goes to some trouble to posit a mechanism for the restoration of order from a universe, or at least a solar system, that has gone to its final rest. While many "aggregates of matter" within the universe "are passing through those stages—which must end in local rest, there are others which, having barely commenced the series of changes constituting Evolution, are on the way to become theatres of life." Spencer further allows for the possibility of an ultimate renewal by extending the implications of local renewal: "Certain of the great facts which science has established imply potential renewals of life, now in one region now in another, followed, possibly, at a period unimaginably remote by a more general renewal."[76] If life may be restored here and there, perhaps it may be restored everywhere. It is a dubious hope, since his "moving equilibrium" requires some external energy source. How do we get outside the universe? But even William Thomson famously allocates to the "great storehouse of creation" the potential for universal renewal currently unimaginable to us—a possibility that Rutherford turns into prophecy with the discovery of nuclear fission.[77]

But while Spencer's must forever remain a speculative answer to a speculative inquiry,[78] Dickens signals a way out of the naturalist bind because he chooses to concern himself with the individual and the family rather than the universe. Enacting not so much the "conflict between . . . the progressive vision of Darwinism and the degenerative vision of thermodynamics,"[79] but the progress allowable within thermodynamics itself, *A Tale of Two Cities* finally evinces a qualified thermodynamic optimism. For Spencer, Darwinian evolution is merely one instance of a more general evolution, itself simply one direction in which a universe governed by the persistence of force may tend. For *A Tale* too, the persistence of force accounts for progress as well as degradation, the large as well as the small. The relation between these makes possible "a new kind of heroism." Small systems, at least, may resist the larger entropic trends, restoring personal energies and securing domestic order: "Carton's act makes possible the domestic norm, yet is entirely outside that norm. Violent death is a condition of order and domesticity."[80] Order can nowhere be increased, unless we pay the price of disorder elsewhere. Entropy is the left hand of work. And even in a universe driving to heat death, we may make a little order. It is a modest hope, perhaps, but one that may sustain even the bleakest of houses.

III The Engine and the Demon: Transformations

5 *Bleak House*: The Novel as Engine

We may regard the Universe in the light of a vast physical machine.
—Balfour Stewart[1]

In this house, we obey the laws of thermodynamics.
—Homer Simpson (on Lisa's discovery of a perpetual motion machine)[2]

Engineering Entropy

If the first law of thermodynamics was born of religious conviction,[3] the second was the discovery of frustrated engineers. Joule's experiments, Kelvin's articulations, Clausius's nomenclature—all of these attempt to account for the heat loss driving engineers a little batty since before the time of Sadi Carnot. Engineers struggled to make more efficient engines; heat loss kept getting in the way. One could put fuel in and get work out, but always less than was hoped for or even expected and never reversibly. But not for lack of trying. And the trying, after all, turns out to be very important. Sydney Carton, as we have seen, quite effectively manages to produce more work than was expected. He is Dickens's tool for producing novelistic order within the constraints of thermodynamics—for manufacturing what some like to call a happy ending. But Carton is not the production of a moment. By the time Dickens wrote *A Tale of Two Cities*, he had already done quite a bit of tinkering with producing order in spite of entropy. In Dickens's *Bleak House* we find evidence of this tinkering. For while *A Tale* gestures back toward an earlier time and physical sensibility, mostly critically (as in the case of Monseigneur) but also with some nostalgia (as with Mr. Lorry's watch), *Bleak House* was actually published about seven years earlier.[4] And in spite of Dickens's well-known, if qualified optimism, one might reasonably expect *Bleak House* to be bleak. Nor does the novel disappoint. It does, in fact, manifest a profound sense of entropic decay. Like so many of the decadent texts of the late century (more on these in the next chapter), it seems almost overwhelmed by entropy. How, we may reasonably ask, is this possible? After all, the term *entropy* was a decade away. But engineers were struggling to make better engines long before the scientific object *entropy* came into being. And like them, *Bleak House* struggles with the list of features that will become entropy—the tendency of all systems to decay, the difficulty of maintaining order or producing work.[5] But at the moment of its emergence, the new scientific object has "*no other shape than this list*" of answers to laboratory trials, this list of what it does when you do stuff to it. As Bruno Latour says, "If you add an item to this list, you *redefine the object*, that is, you give it a new shape."[6] *Energy* is precisely the list of things it can do: the things it turns into, the fact that it is always conserved. Something that could be suddenly created or destroyed by "man's agency" alone would not be the scientific object we have come to know and love as energy. Similarly, if it decreases spontaneously (without, say, someone putting work into the system), I don't know what it is, but it's not entropy.

As fiction, like engineering, searched for physically allowable alternatives to an apparently inevitable, scientifically predicted decay, *Bleak House*, for all of its bleakness, engaged in the search. This long and complicated novel (a miniseries just waiting to happen) proves deeply thermopoetic; it is littered with incomplete experiments and partially successful prototypes, with fictional systems that struggle to produce order with varied and generally limited success. This chapter considers Dickens as a novelistic engineer; *Bleak House*, as a work of extensive novelistic experimentation in the field of applied energy physics. As Dickens explores the system constraints that will come to be known as *entropy*, he tests and rejects various models for textual/ interpersonal engines, building prototype after prototype in his quest to figure out what works and how. He finds, as it happens, results akin to those of James Watt. So those of us who have never heard of a "Carnot cycle" will learn a few things about how engines work, about sources and sinks, generators and condensers, even about James Watt and the Newcomen steam pump. And for new fans of *Bleak House* as well as old, the novel itself comes to look like an engine. It takes on a distinctly thermopoetic structure, even as its overwhelmingly entropic vision is mitigated by practical considerations, which reduce entropy from the herald of universal decay to a mere complication, however intractable, in the building of better engines.[7]

Exploding Facts

With *Bleak House*, we also have another opportunity to see how fiction participates in the process of making scientific facts. Entropy in particular is sent further downstream in the direction of fact; its contours—those things it does—look increasingly like those that define the late nineteenth-century object we call by that name. Of course, it is necessary for Dickens and for entropy that the universe thus cooperates. Fact making is neither a small nor an isolated task, but rather one that involves a complex system of alliances between human and nonhuman actors. Moreover, to send the list of features—universal decay and so on—downstream in the direction of fact, to develop the scientific object that will be called entropy, cannot be the work of fiction or language alone. To illustrate and emphasize, it helps to look at one attempt at fact making that is somewhat less than successful.

One group of readers, on hearing that there would be talk of physics and *Bleak House*, sat upright, blinked twice in sudden recognition, and exclaimed, "Oh yes! That spontaneous combustion thing." Another group, possibly overlapping the first, is now thinking, "Spontaneous combustion! That's not a scientific fact." Ever diplomatic, I interject, "You're both right." On the one hand, Dickens's inclusion of the phenomenon—which he stages earnestly, exploding an aptly named Mr. Krook—gives us an important clue to his investment in physical science and his connections to Michael Faraday.[8] On the other hand, spontaneous combustion does not survive the subsequent scientific trials to which it is put. The universe (my shorthand for a variety of human and nonhuman actors) does not cooperate. And without this cooperation, Dickens cannot make a fact out of spontaneous combustion.

Lucky for us. Not only because we and our loved ones remain safe from suddenly bursting into flames, but also because the wannabe fact brings Dickens's investment in fact making into sharp relief. Identifying spontaneous combustion as "the death of all lord chancellors in all courts and of all authorities in all places," Dickens finds its trigger "in the corrupted humours of the vicious body itself."[9] This in itself seems fanciful enough, but Dickens signals his investment in fact making in the preface to the first edition, where he promises that he does not "wilfully or negligently mislead [his] readers" regarding spontaneous combustion. Having investigated the subject before writing of it, he insists that there are "about thirty cases on record." In the preface, he alludes to "the recorded opinions and experiences of distinguished medical professors, French, English, and Scotch" and even mentions the name of his "good friend Mr. [George Henry] Lewes," who seems to have reversed his opinion—not "that spontaneous combustion could not possibly be" (that one he stuck to), but "in supposing the thing to have been abandoned by all authorities."[10] When Lewes originally challenged the episode, Dickens responded in the next installment of *Bleak House*, incorporating in the text proper, a set of gestures like those that eventually grace the preface. Among "learned talk of inflammable gases and phosphuretted hydrogen," he reminds us of a variety of "other authorities," whose experiences should, he suggests, trump the implications of any scientific theories.[11]

Lewes has quite a bit to say on these so-called authorities,[12] but the effect of such gestures in the text is double-edged. First, this is a blatant attempt to "factify" spontaneous combustion. Dickens's novel pulls several moves familiar to that other rhetorical vehicle, the scientific paper. Through the argument from authority and the piling up of references, he presents spontaneous combustion as a fact that has already undergone many trials, musters his many allies, suggests just how many people—some of them quite learned and trustworthy—one would have to doubt if they doubt this.[13] Even Lewes acknowledges the potential effectiveness of such efforts, worrying that the

"magnificent popularity" of his dear Dickens "will tend to perpetuate the error in spite of the labours of a thousand philosophers."[14] There is another side to Dickens's inclusions, however,
that might have allayed Lewes's fears. For even as Dickens presents spontaneous combustion as fact, he reminds us that it is
under contention. Paradoxically, the move to win the controversy and establish the fact is also the move that reminds us
that "spontaneous combustion is possible" is not a fact at all,
but a controversial statement. Dickens's allusion to a variety of
people-who-said-so reminds us of the source of statements like
this one, and of course, gives Lewes the opportunity to object
to these sources as well as to the strategy of arguing from
authority. In this way, the allusion to the controversy evoked
by Krook's death—both in the story and among its readers—
pushes spontaneous combustion upstream, as we have, in the
direction of artifact, where it will eventually settle down for a
good long time.

Entropy Downstream

This, as we know, is not the fate of entropy or energy or the
various principles of engine design, which in *Bleak House* operate
so tacitly as fact that we may fail to notice them. *Bleak House*
proves deeply invested in the broader thermopoetic conversation, and without once using the word—how could he?—
Dickens nonetheless works to factify entropy itself. For him,
it functions as a fundamental premise on which he builds his
more complex and less established claims. Indeed, when it
comes to entropy, Dickens wears his mind on his sleeve. Saving
us the trouble of scrounging for evidence that he is thinking
about universal decay, he announces it on the first page. Indeed,

this novel opens with an image of local decay that evokes the end of life itself. Smoke rains from chimney pots, "with flakes of soot in it as big as full-grown snowflakes—gone into mourning, one might imagine, for the death of the sun."[15] Dickens's black snow not only anticipates images of darkness and cold we will soon see in Flammarion and Wells,[16] but also reminds us that poets (and lunatics, of course) have always worried that the sun would go out. Indeed, when Tennyson dreamt that "Nature's ancient power was lost," he too imagined the streets "black with smoke and frost."[17] In Dickens, the sun itself has gone out; the city—like its dogs—is mired in indistinction. And on a smaller scale, the fires of London prove similarly unable to provide heat or light, order or distinction. Throughout, we see further evidence of heat death, of energy sources running down: darkness, cold, fatigue. Shops are lit two hours early, presumably because of the lack of sunlight, and "gas loom[s] through the fog in divers places in the streets, much as the sun may." The gas itself "has a haggard and unwilling look."[18] It perhaps goes without saying that such haggard gas lamps are no substitute for the absent sun, which even if not truly dead, certainly does not shine here: "The streets were so full of dense brown smoke that scarcely anything was to be seen." The fog suggests the absence of any adequate source of heat and light, for when our heroine and sometimes narrator, Esther Summerson, first arrives, she mistakes this "London particular" for evidence of a "great fire."[19]

Dimness also pervades the Court of Chancery, which shows signs of the same decay that plagues London.[20] As we enter the dim court, we experience more of Dickens's thermodynamic sensibilities. Its "wasting candles," the fog that hangs heavy within, and "the stained-glass windows [that] lose their colour and admit no light of day into the place" all provide evidence of decay. The phrase "wasting candles" implicates humans who would waste candles, even as it suggests the thermodynamic inevitability that those candles will waste themselves with or without our help. In Chancery, moreover, Dickens ties the burning out of energy sources to the only somewhat more figurative burning out of individuals. Even the court personnel evince a haggard and rundown look, "all yawning, for no crumb of amusement ever falls from Jarndyce and Jarndyce (the cause in hand), which was squeezed dry years upon years ago." In the realm of human affairs, the drive to disorder is as palpable as the darkness. Those who would survive in Chancery must accept, even embrace, the increase of entropy, which dominates all of its processes. "In the midst of the mud and at the heart of the fog" where the Lord High Chancellor sits, most have a "loose way of letting bad things alone to take their own bad course, and a loose belief that if the world go wrong, it was, in some offhand manner, never meant to go right."[21] And indeed, according to the second law of thermodynamics, it never was.

Heat Sinks, or Abandon Energy All Ye Who Enter Here

Entropy manifests itself in black snow, in the failure of light, warmth, or color, as well as in boredom and fatigue. It also manifests itself as heat. "There never was such an infernal cauldron as that Chancery on the face of the earth!" exclaims the energetic Mr. Boythorn (more on him later), identifying Chancery with heat dedicated to melting, to the production of indistinction. In fact, he understands Chancery as so given over to entropic decay that we can do nothing more than accelerate the process, to have "the whole blown to atoms with ten thousand hundredweight of gunpowder."[22] All we can do is dedicate the stored chemical energy of all that gunpowder to the production of atoms in random motion. All we can do with all that heat is produce more heat. Thus *Bleak House* proves big enough to stage the central paradox of heat death—that entropic decay manifests itself not only as cold, but also as what we call heat. It is the fate of large systems that cold and heat are subject to such indistinction; they are the same thing—the random motion of molecules—when there is nothing outside to compare them to. And Chancery, like London and the universe, is simply so big that we can't get outside.

Chancery thus functions as an enormous heat sink, sucking up the energies of those who venture into too close contact and causing the deaths of Tom Jarndyce, Richard Carstone, and Mr. Gridley. Tom Jarndyce, the legendary initiator of Jarndyce and Jarndyce, likened Chancery to "being ground to bits in a slow mill" or "being roasted at a slow fire" before taking a gun to his head.[23] Richard Carstone is our tragic hero whose involvement in the case will be his undoing. But Mr. Gridley, "a man of a robust will, and surprising energy," seems to have no other function than to enact the draining of the energetic individual by the entropic system. The parasitic Mr. Skimpole imagines Gridley wandering about in search of "something to expend his superfluous combativeness upon . . . when the Court of Chancery came in his way, and accommodated him with the exact thing he wanted,"[24] that is, something to suck up all that extra energy. As Gridley lies dying, the clever and energetic detective, Mr. Bucket, reiterates Skimpole's characterization. Attempting to buck up his old acquaintance, he expresses surprise "to hear a man of your energy talk of giving in." "You mustn't do that," he continues; "You're half the fun of the fair in the Court of Chancery."[25] But (unlike Bucket, who seems to be able to hold more than others) Gridley is already squeezed dry. Chancery provides a sink that absorbs his superfluous energy and more.

But a heat sink needn't be large. Just as we have often seen the principles of energy reiterated on vastly different scales, *Bleak House* gives us Chancery in miniature, for the combustible Mr. Krook is christened "Chancery" by his neighbors for his simultaneously absorptive and entropic qualities. "Wasting away and going to rack and ruin," he reproduces in miniature the function of the larger energy sink. At his establishment, "everything seemed to be bought and nothing to be sold." And Krook himself sees to it that what goes in does not come out: "All's fish that comes to my net. And I can't abear to part with anything I once lay hold of." Moreover, energy is absorbed without producing any work, since neither "sweeping, nor scouring, nor cleaning, nor repairing [goes] on" about Krook's.[26] Like Chancery, he is a system driving so fast to entropy that nothing is to be done with him except to blow him up. And if Dickens's choice of spontaneous combustion to serve this narrative purpose is not physically possible, it nonetheless represents the problem of an excess of heat reasonably well.

Krook's, moreover, is merely one example of the heat sink localized. That Master of Deportment, Mr. Turveydrop, is another such energy sink, and one moreover that suggests two specific problems in energetics: those pertaining to domestic arrangements and those pertaining to class relations. I will consider both of these in greater detail, but first, the absorptive Mr. Turveydrop (if you please), whose renowned "deportment" seems a far cry from entropy. But Turveydrop is explicit on the decay sustained by the class of gentleman, even to the nation itself: "England—alas, my country!—has degenerated very much, and is degenerating every day. She has not many gentlemen left. We are few. I see nothing to succeed us but a race of weavers." Though some of us may understand the succeeding race of weavers as far more capable of useful work, Turveydrop laments the decline he perceives in the loss of class distinction: "We have degenerated. . . . A levelling age is not favourable to deportment." This leveling echoes the loss of class distinction that Spencer associates with "disorderly homogeneity"—the kind of equilibrium that disables work. But Turveydrop manages to draw energies to himself, and to have considerable work done on his behalf. "In spite of the man's absorbing selfishness, his wife (overpowered by his Deportment)" believes in him to the last, and impresses on their son Prince his father's "inextinguishable claim upon him."[27] Turveydrop's selfishness absorbs; his deportment overpowers. The processes are almost the same; they use up his wife in their sustenance and move to absorbing his son. Soon, his daughter-in-law Caddy (née Jellyby) will devote her energies to the sustenance of his deportment, and even her baby—tiny and immobile—will reveal the effects of this drain all too well.

Esther Summerson, however, identifies Prince Turveydrop as "indefatigable."[28] For all that Turveydrop acts as a drain on his energies and Caddy's, both manage not only to sustain his deportment, but to do additional work besides. There remains usable energy in the world. In spite of the broad and inevitable tendency to entropy evinced by every large and many a small system, solar to social, there are opportunities to do useful work and to decrease entropy somewhere. Caddy's baby is born, her household established, their dance school run. More broadly, Bleak House is renovated, Rosa married to Watt, Mr. George saved, Ada recovered, and many more babies born. How is this possible? Surrounded by Chancery, London, the universe, and having in residence absorptive individuals like Krook or Mr. Turveydrop or Mr. Skimpole just sucking up energy like it's going out of style (which of course it is), how are we to get anything done?

Well, it helps if we take a moment to think statistically. In the billions and billions of interchanges between, say, the molecules of a gas, some will gain and some will lose. Two molecules collide in a way that we can't help but picture as two billiard balls; one speeds up, the other slows down. If energy is conserved overall, we are still within the bounds of law. But the energy of one has increased at the expense of the other. Entropy teaches us that it is a less-than-zero-sum game, but here and there, someone wins.

Directing Energy

In the millions of collisions in the social gas of *Bleak House*, the poor boy who shivers in the crossing seems to turn up an unquestioned loser. His name is Jo. Though "not quite in outer darkness" and sustained by a "distant ray of light,"[29] he is deprived of his best source of energy and hope. His energies are used as information is drawn from him as a witness to the life (though not the death) of the Mr. Hawdon who was good to him.[30] They are wasted as he is rejected as a witness and driven out of town. But Skimpole, who turns out to be a particularly savvy character on questions of energy, suggests the possibility that Jo may have had, may have used energy to produce something—if not socially useful work, which is of little interest to Skimpole, then at least poetry. Infuriatingly unsympathetic,

Skimpole would have been better pleased if Jo had shown "some misdirected energy that got him into prison. There would be more of an adventurous spirit in it, and consequently more of a certain sort of poetry." In a world so broadly characterized by the dissipation of energy without the accomplishment of any work, perhaps doing anything is to be admired. Certainly Skimpole, who so studiously avoids doing anything himself, admires and thrives by the work of others. And where the law is so thoroughly a drain of useful energies, breaking it seems like an accomplishment indeed. As Skimpole further elaborates his opinions regarding Jo, he identifies society's failure to serve as energy source rather than sink:

> At our young friend's natural dinner hour, most likely about noon, our young friend says in effect to society, "I am hungry; will you have the goodness to produce your spoon, and feed me?" Society, which has taken upon itself the general arrangement of the whole system of spoons, and professes to have a spoon for our young friend, does *not* produce that spoon; and our young friend, therefore, says "You really must excuse me if I seize it." Now, this appears to me a case of *misdirected energy*, which has a certain amount of reason in it and a certain amount of romance; and I don't know but what I should be more interested in our young friend, as an illustration of such a case, than merely as a poor vagabond—which any one can be. [31]

Though we can't allow ourselves to believe that Skimpole would summon any sympathy for Jo, had he stolen food and gone to prison, we can readily believe that Skimpole would better identify with that decision. For us, however, it is important that he makes the distinction between doing something (seizing a spoon) and doing nothing (being a vagabond), between misdirecting energy and showing no particular signs of having any to use. The act of seizing a spoon would suggest not only that usable energy exists (the spoon serving presumably as its conveyance), but also that one can draw it to oneself. We can even imagine that the acquisition of food would enable Jo to direct the energy thus acquired into other work, beyond even the illustration of reason and romance. He has already proven useful in providing information to Lady Dedlock (on whom, more later). Who knows what he may have accomplished on a reasonable diet?

Meanwhile, I use poor Jo once again, this time to illustrate the relation between individual and system. For if he were to "misdirect his energy" in the seizing of a spoon, he would draw usable energy to himself. He would be thrown into prison, however, for producing disorder in the "general arrangement of the whole system of spoons." And this simultaneous production of local order at the expense of systemic disorder is an object of much interest to the text broadly. As we work our way up to engine design, Jo's seizing a spoon serves as a hypothetical example of what we will call our first principle, the principle that makes work possible: *You can do work, even increase order here, but you will pay for it in increased disorder (eventually entropy) somewhere else.* Jo's hypothesized *mis*directing energy strongly implies the possibility of his directing it well. Perhaps with more or better energy to begin with, perhaps with better means of directing it, others may accomplish considerable work.

Jo's tragedy, however, anticipates Richard Carstone's. And this tragedy is increased because of the type and quantity of Richard's energy: "I am young and earnest," he announces, "and energy and determination have done wonders many a time. Others have only half thrown themselves into it. I devote myself to it. I make it the object of my life." Claiming to have that much more direction than others, Richard predicts a better outcome for the use of his energies. And in fact (after several false starts), he simulates true direction, resisting that which might drain energy from his purpose, even declining a visit to Lincolnshire and "making most energetic attempts to unravel the mysteries of the fatal suit." Richard's case, moreover, raises the question of whether what matters is the kind of energy or its direction. We are told that his "energy was of . . . an impatient and fitful kind." But what does this imply? Is his energy potentially useful but misdirected, or is there something about his particular energy that suggests that it can never have done useful work? As he directs his energies toward Jarndyce and Jarndyce, his new "unvarying purpose in life," we cannot determine whether the case effectively draws to waste what might have been useful or only gathers energy that has no direction of its own.[32] Much of *Bleak House*, however, suggests that various forms of energy, here as elsewhere, prove interconvertible, that we can in fact choose how to direct our energies, and that choice makes all the difference.

Inspector Bucket, for example, seems to manage his relation with the system remarkably well. A man of considerable personal energy, Bucket is the object of Esther's wonder, as she observes that "the energy of my companion never slackened." Even when, in their pursuit of Lady Dedlock, the horses "had stopped exhausted half-way up hills . . . he and his little lantern had been always ready."[33] Armed with a light that never goes out, Bucket too seems never to wane. He even seems impervious to draining by the systems around him. He is enervated neither by his contact with Chancery nor by that with the Dedlocks. Bucket's supply remains always intact and apparently untapped. But if Esther reminds us of what we already know—Bucket is a man of energy—Skimpole's description ties him to Jo. For Skimpole finds this "active police-officer" to be "a person of a peculiarly directed energy and great subtlety both of conception and execution."[34] Skimpole, of course, wishes to draw from Bucket as source, in other words, to take a bribe. But in identifying Bucket as a person of "peculiarly directed energy" Skimpole suggests that it is not just strangely directed energy but particularly, highly, or well-directed energy. It must be the latter in order to accomplish useful work. Skimpole thus suggests that direction is key—not just having energy, but how one uses it.

Each of these examples suggests, in a qualified way, the presence of usable energy and the potential for accomplishing work, but Jo's case has raised the question of where it comes from. Local order, we assume, has a cost. This suggestion regarding Jo raises an important question regarding Bucket: to what extent does the production of order that he effects produce disorder elsewhere in the large system in which he works? The question, here reframed in energetic terms, is familiar: Is Bucket, "who discovers our friends and enemies for us when they run away, recovers our property for us when we are robbed, avenges us comfortably when we are murdered,"[35] complicit in the disorder he works against? The answer, energetically speaking, is that he must be. Practically speaking, Bucket commands our attention; somehow he has energy or draws it from the larger system in order to accomplish useful work. Bucket is an engine.

What Work Is

What then, is an engine?[36] Broadly speaking, anything that uses energy is an engine. And vice versa. If we wish to be slightly more specific, we need only emphasize *uses*. Anything that uses energy to do something—work—is an engine. All the rest is commentary. Dickens's social engines seem to work like simple heat engines, having (as we are beginning to see) both a source (or sources) and a sink (or heat bath) and, with a little luck, doing some work. So far, I haven't troubled too much about the specifics of work, but as it happens *work*, like *force* and *energy*, is an evolving term. Work gets more and more attention as the nineteenth century progresses: "More than an occupation, a means of earning a living, or even an indicator of social status, work was increasingly regarded as a moral imperative as much as a physical necessity." Even in these senses, *work* resonates with physical-science possibilities, functioning as "a *measure* of a person's moral worth in just the same way as it was an *indicator* of an individual's economic *value*."[37]

Work also comes to have a specific definition in engineering and physics. Synonymous with *mechanical effect* or the *force applied over some distance*, another definition of work falls out of the simple algebra of heat engines. It will be convenient at this point to introduce a simple equation. And just in case we ever run into any engineers, we are going to stick with convention and call heat Q. For this and more obvious reasons, we are going to call work W. Okay, here goes:

$$Q_H = W + Q_C \tag{1}$$

What we have here is a simple but profound statement of where the energy goes. We also have a neat wrap-up of some of the elements of engine design we have been working on: sources, sinks, and work. Whatever energy (Q_H) goes into the engine from the hot spot or source yields some work (W) and some waste heat (Q_C), which just warms up the cold spot or sink. Here's the rub: Q_C is always greater than zero. This is, by the way, another way to state our first principle: *We can convert energy to work, but never perfectly*. We always get out less than we put in. Or as my high school physics teacher used to say, "You can't get ahead; you can't even break even." Some of that nice usable energy (Q_H) will go to warming the cold spot or sink, which thus earns the somewhat counterintuitive moniker of *heat bath*. So, as we do work, we also end up increasing the entropy in the surrounding system, to warming (however slightly) Chancery, London, and the universe. And we can never ever get it back. What's worse, as I also learned in high school, "You can't get out of the game." That is, all these rules hold as long as we are in a closed system.

The good news is in the *W*. Here's another definition of an engine, a particularly optimistic one: If all the energy in the universe will eventually convert to heat anyway, the job of an engine is to get in the way. An engine diverts energy on its path to heat. Work is a transitional state; it is a form of energy, a form we like. It can even be converted back to Q_H and reused, but eventually it must end up as Q_C. Meanwhile, we may as well enjoy it. Or at any rate, do something useful with it.

Problematic Prototypes, or Trials and Errors

What constitutes "something useful," however, is up to us. *Bleak House* is deep in the business of satirizing those who insist they are doing useful work, when they are in fact doing harm. And not just the lawyers either. Almost as bad are the philanthropists. With all the harm done, all the waste and disorder produced by Chancery, we hardly need our two doubtful philanthropists, Mrs. Jellyby and Mrs. Pardiggle, to reiterate the point. But these serve as important prototypes. They are models of engines. And if they don't seem particularly useful, they nonetheless produce a lot of work.

In striking contrast to such individuals as Mr. Gridley and Richard Carstone, Mrs. Jellyby and Mrs. Pardiggle seem to thrive in spite of the entropy that surrounds them. One might go so far as to say they thrive because of it. And the energy they enjoy is in the form of work: "The prominent point in my character," Mrs. Pardiggle tells us repeatedly, is that "I love hard work; I enjoy hard work. . . . I don't know what fatigue is." Her energy, moreover, never seems to run out. In her family and her philanthropy, she claims to be immune to the effects of either individual or environmental energy sinks. To the needy family she has purportedly come to help, she asserts, "You can't tire me, good people." Her own family she informs, "I do not understand what it is to be tired; you cannot tire me if you try!" Confronting each domestic scene before her as if it were a sink designed to drain her energies, she manages rather to draw from it, claiming to enjoy the hard work and undoubtedly thriving.[38]

That she manages to draw energy from such stores requires a facility with energy conversion of no mean order. But though she may seem an energetic individual, Mrs. Pardiggle can't properly qualify as an energy source. We know the sources of her vast supplies of energy; she draws them from her young family and from Mr. Pardiggle, who have been "quite worn out with witnessing" the astonishing "quantity of exertion (which is no exertion to [her])" that she goes through, though she herself remains "fresh as a lark." Her capacity to draw such energy from others and convert it into work accounts for what is aptly termed her "mechanical way of taking possession of people."[39] And the people thus taken possession of, cannot help but be worse for the association. For work costs. For every W, a little Q_C must fall. Or a lot. Encounters with Mrs. Pardiggle render the

victims of her philanthropy among the saddest cases of depletion and dissipation that *Bleak House* has to offer: "a woman
with a black eye, nursing a poor little gasping baby by the fire;
a man . . . looking very dissipated, lying at full length on the
ground, smoking a pipe."[40] Certainly, in need of real aid, the
woman, the gasping baby, and the prostrate man know what
fatigue is. Indeed, the striking use of the term *dissipated*—the
only such use in *Bleak House*—resonates with the social and
physical significance we have seen before.

What the dissipation of this poor family as well as of the
Pardiggles suggests, Mrs. Jellyby's family announces in no uncertain terms. Like Mrs. Pardiggle, Mrs. Jellyby seems well endowed
with energy, a woman of "immense power."[41] Roughly synonymous with energy, force, and momentum as well as strength,
authority, and influence, *power* has a specific physical definition
as well. It refers to the rate at which work is done, energy trans-
ferred or transformed. And Mrs. Jellyby's energies (rapidly trans-
ferred) are garnered at the expense of her family. We find out
parenthetically what is emblematic of the whole household:
"The fire had gone out, and there was nothing in the grate but
ashes, a bundle of wood, and a poker." The house is cold; the
food, uncooked; the disorder, of proverbial proportions. The
family itself is described by Miss Caddy Jellyby as "nothing but
bills, dirt, waste, noise, tumbles downstairs, confusion, and
wretchedness." The disorder and the failure of work are, moreover, all of a piece. For Mr. Jellyby, too, is "merged—in the more
shining qualities of his wife," his "scrambling home, from
week's end to week's end . . . like one great washing-day—only
nothing's washed!"[42] Caddy herself is similarly drained of energy
to no apparent end but increased disorder. When she appears
later in Esther's room "shivering there with a broken candle in

a broken candlestick in one hand and an egg-cup in the other,"[43] we get the message. Mrs. Jellyby—who on her first appearance significantly snuffs two candles—is guilty of snuffing or draining the sources of heat and light around her and of rendering attempts at domestic work (even the proper use of an egg cup) futile.

The imagery of the candle, moreover, bespeaks Dickens's particular connection to thermodynamics—his interest in Faraday, his request for the famous scientist's lecture notes on "The Chemical History of a Candle," and his role in the subsequent revision of these notes for *Household Words*. But as young Harry Wilkinson informs us in that comical chemical history, the candle does not burn to nothing, "everything . . . goes somewhere."[44] All of Caddy's work—for she complains that she "can't do anything hardly, except write"—is directed by Mrs. Jellyby toward Africa. Indeed, all of the energy thus garnered, the energies Mrs. Jellyby will call her own, go into her pet project, the cultivation of both coffee and natives in Borrioboola-Gha: "It involves the devotion of all my energies, such as they are; but that is nothing." She can, however, dismiss the contribution of *her* energies as unimportant because they are indeed "nothing." They are not hers. Like Mrs. Pardiggle, she draws them from her miserable, drained, and disorderly family. We can hardly help but respond to Mrs. Jellyby's complaint that "my work is never done"[45] with a somewhat caustic "No kidding." But of course, she means her "telescopic philanthropy." The mismatch between her intentions and the way her claim sounds to anyone with the most passing familiarity with her housekeeping, suggests that useful work—mechanical or social—may be hard to identify. *Whether usable energy is well used as it is converted to work clearly depends on how such work is directed*—our second principle. Mrs. Jellyby, *Bleak House* suggests, does not direct hers very well. For not only does telescopic philanthropy take place at the expense of local order, Africa itself is particularly large and very far away. The family is simply too small a source; Africa, too large a sink for the work ever to be done. Mrs. Jellyby thus suggests (by negative example) that if we really want to get some work done, we ought to choose a smaller sink.

A Better Engine

Bleak House does, in fact, present us with a few domestic engines that seem to work. Mrs. Bagnet runs one, and "the great delight and energy" her daughters exert in their duties suggest that they will as well.[46] Of course none is so effective as Esther Summerson. We are introduced to her in a chapter called "A Progress," which suggests how much her ability to do useful work bucks the larger entropic trends of the novel. Her name, evoking both a star and a season, suggests her close association with light and heat. And her first-person narration makes any assessment of her beauty secondary; what draws so many to her is her warmth. She seems to be a source of usable energy.[47] Of course, what we must mean by this is not that she creates energy—that wouldn't be consistent with the first law—but that she stores it or converts it into useful forms. She often repeats her childhood vow to be "industrious . . . and to do some good to some one"—to do, from an energetic perspective, useful work. She reiterates this vow in her intention to be "dreadfully industrious" when she feels a drain on her own spirits.[48] Why this industry should be dreadful evokes the larger textual anxiety regarding women's energies. But for now, the effectiveness of Esther's domestic energies speaks to the momentarily more pressing question of engine design.

Temporarily established at the Jellybys, Esther begins her assault on the entropy of the environment, "making our room a little tidy, and . . . coaxing a very cross fire that had been lighted, to burn; which at last it did, quite brightly."[49] The ordering she does here anticipates her work as the housekeeper of Bleak House and as sometimes-narrator, in which roles she lends her ordering capacity to the house and to the novel.[50] An effective engine, she does real work even against an overwhelming drive to disorder. Why, then, do some engines work better than others? In stark contrast with Mrs. Jellyby's work, Esther's mechanical work of moving things around actually raises the level of order in the environment, at once tidying the room and creating a usable hot spot, for Esther lights a fire. Both the brightly burning fire and the warm room increase the thermodynamic order of the household. They make a difference—a temperature difference. And this is what makes an engine run. Thus Esther's case suggests our third principle of engine design: *work requires distinction*. Remember equation 1? Well, equation 1 tells us that whenever $Q_H = Q_C$, then $W = 0$. If the temperature of the hot spot is the same as the temperature of the cold spot, we can't get any work out of the system.

This principle has governed engines forever. While the scientific object *energy* is still emerging, this principle appears as a "distinct analogy between Sadi Carnot's model of caloric doing work by flowing from one temperature level to another lower one and his father Lazare's analysis of the work done by water turning a waterwheel while flowing from a higher level to a lower."[51] There must always be a higher energy level and a lower one, for work is produced in the fall. In the case of a waterwheel, the fall is literal. In a water mill, the work done is the processing of cotton into cloth. In the case of a turbine generator, the mechanical energy of falling water is converted into usable energy of another form, electricity. Water seeks its own level; the fall happens spontaneously. The engine gets in the way. But what if the water has already found its own level, so to speak? Not much. If we want our engine to keep working, we must maintain its operative distinction.

Esther's case suggests one way of doing so. Esther's work is successful because she chooses to help in small ways. Never having "turned [her] thoughts to Africa," she focuses rather on smaller systems—an individual or family she can serve—and with far greater success. Very Dickensian, one might say. But also thermodynamically savvy. Esther's may seem like a small contribution, but that is its genius. If Mrs. Jellyby looks "down upon [her] rather for being so frivolous" —that is, for failing to attend to a philanthropic conversation, in favor of attending to the neglected Jellyby children—we know Esther's secret is in having "no higher pretensions."[52] Because, as it happens, size does matter. Though the laws of energy apply whether one is dealing with a big system or little system, when one is thinking about the interaction between systems, their practical consequences may vary. And one might do better—from the perspective of making use of available energy—to deal with a heat sink of one's own size.

Which brings us to literal heat engines—or more specifically, steam engines. The early eighteenth century saw the invention, by Thomas Newcomen, of a steam engine capable of producing mechanical work. It worked something like this: A hot spot, a boiler, produces steam. Hot steam cools—as is usual in this lukewarm world of ours—and condenses as water. But place it in a well-sealed container, and it creates a vacuum as it does so. This is because water in liquid form occupies less space than steam. If air can't get in to occupy the extra space, a vacuum—

truly empty space—is created. (Actually, what you get is a partial vacuum or low-pressure area.) But nature, as we have heard, abhors a vacuum. Space would much rather be filled. So the vacuum tends to draw stuff into it, to devacuumize. In the case of Newcomen's pump, the vacuum was used to draw water—to exert force over a distance, to do work. However, in Newcomen's engine, the heated steam cooled simply because everything around it—the engine itself, the air, whatever—was relatively cool. Steam condensed everywhere. The problem was this: not only did a lot of otherwise usable energy get spent in warming up the world, the cold parts of the engine also warmed up. The hot spots and cold spots grew closer in temperature. As the distinction so necessary to its proper functioning diminished, the Newcomen engine worked less and less well.

Toward the end of the eighteenth century, James Watt effectively fixed the problem. His name has thus "become synonymous with 'steam.'" As the unit in which we measure power, it now, as one historian so eloquently puts it, "glows discreetly in every electrical light."[53] Anyway in a very small nutshell, James Watt decided after much tinkering that it wasn't so much a matter of keeping the hot spots hot; that was a relatively easy task if one only had enough fuel to burn. It was a matter of keeping the cold spots cold. Somehow, he needed to condense the steam without warming up the whole mechanism. So he introduced the *condenser*—a mechanism for drawing away condensed steam and coolant water from the other parts of the engine. Like Dickens, he uses a relatively small, localized heat sink. This brings us to what we might count as our fourth principle of engine design, though it is closely related to the second: *Keep things in proportion.* (That seems generally good advice.) Don't heat up the whole world. (That suggests an eco-conscience.) To better direct where the energy goes, get a small sink—a condenser.

A Move in the James Watt Direction

This little digression into the life of steam engines thus returns us (perhaps breathing a sigh of relief) to Dickens. Characters like Bucket manage work in spite of their contact with the novel's biggest heat sinks: the Dedlocks, Chancery, London. They are, arguably, the textual equivalent of the Newcomen engine. Invested with energy, they accomplish something as it dissipates into the larger system, though it may be not much more than the slowing of dissipation or the maintenance of the current state of order. But *Bleak House* pursues its experiments, performing adjustments akin to Watt's, tweaking sources and sinks, striving to direct the flow of energy in the search for a more efficient engine.

There can be no doubt, moreover, that Dickens—himself "associated with the Mechanics' Institution named after James Watt"[54]—has Watt's engine in mind. His qualified homage to the famous inventor appears in the form of his man of industry, Mr. Rouncewell, who, less subtle than Dickens, just names his son Watt. Mr. Rouncewell, it seems, gives up a guaranteed post as steward on the Dedlocks' estate of Chesney Wold to pursue what turns out to be a very successful career in the iron country up north. His story, moreover, recalls the popular stories told about James Watt, who at thirteen was supposed to have been inspired "by the force of steam gushing from a boiling kettle to invent the steam engine."[55] Mr. Rouncewell, as a schoolboy, instead invents a water pump, reiterating the wish to move water about, a wish that had already done so much to drive the development of the steam or fire engine. And though the young Rouncewell's invention is literally for the birds, he tinkers with something that sounds a lot like Newcomen's pump and with a view to improving efficiency, "constructing steam-engines out of saucepans, and setting birds to draw their own water, with the least possible amount of labour." With this "artful contrivance of hydraulic pressure . . . a thirsty canary had only, in a literal sense, to put his shoulder to the wheel, and the job was done."[56] The efficiency of his bird feeder, moreover, anticipates the greater efficiency of his business. Mr. Rouncewell accomplishes work "with the least possible amount of labour." Canaries thus draw water and Mr. Rouncewell draws himself up, becoming a wealthy ironmaster, even securing a seat in Parliament. If Sir Leicester Deadlock bemoans this rise as representing "the obliteration of landmarks, the opening of floodgates, and the uprooting of distinctions," we are not surprised. For when Rouncewell dares to "'draw a parallel between Chesney Wold and . . . a factory'" we understand that the comparison is apt and know why Sir Leicester must "[resist] a disposition to choke."[57] Sir Leicester's effectiveness depends on usable energy he no longer has, on distinctions that he can no longer sustain. Rouncewell's search for Wattlike efficiency, on the other hand, has created new distinctions—hot/cold, high pressure/low pressure, nouveau riche/passé aristocracy—which will do the work of the next century.

The mechanical work of drawing water thus presages the social work of raising its inventor in the world. Indeed, the struggle between Mr. Rouncewell and Sir Leicester Dedlock may remind us of Stewart and Lockyer's man of personal energy and man of position (see chapter 3), whose struggle served as a metaphor for kinetic and potential energy. Mr. Rouncewell embodies the man of action in opposition to Sir Leicester Dedlock's energy of position. And with Sir Leicester, Dickens has already done work to establish the claim reiterated in Stewart and Lockyer, that the man of position is decidedly lacking in personal energy.[58] Moreover, in both texts, what's old is new again. In Stewart and Lockyer, the class struggle is as traditional as throwing stones. In *Bleak House*, this problem of the industrial age recalls uprisings of an agrarian moment. For Rouncewell's "propensity" for mechanical invention seriously unnerves his mother, who "felt it with a mother's anguish to be a move in the Wat Tyler direction." Her thoughts thus turn to a Wat who is not her grandson, but rather the leader of an uprising of medieval workers and one, moreover, whose name had recently been reused by a leader in the Chartist resistance.[59] In this, she echoes Sir Leicester's own "general impression of an aptitude for any art to which smoke and a tall chimney might be considered essential."[60] All her energy bound to the Dedlocks, Mrs. Rouncewell, the housekeeper, is fully committed to the maintenance (not the progress) of Chesney Wold. And her conception of Sir Leicester's objection—to smoke and chimneys—only barely acknowledges the industrial specificity of the threat her son poses to current class boundaries. But the text's punning on *Wat(t)* highlights the modernity that Mrs. Rouncewell downplays. It suggests another interpretation of the same move, fully reinforced in the naming of Rouncewell's most complex product, his son Watt. And though Dickens himself evinces considerable ambivalence about arts involving smoke and tall chimneys, Mr. Rouncewell is allowed to take his place among innovators like James Watt, who helped drive the industrial revolution, and what is broadly construed as its correlative class revolution, by making a more efficient engine.

Maintaining Distinction

Like Chancery, London and the universe, the fashionable world is a large system dominated by the drive to entropy: "Wrapped up in too much jeweller's cotton and fine wool, [this world] cannot hear the rushing of the larger worlds, and cannot see them as they circle round the sun. It is a deadened world, and its growth is sometimes unhealthy for want of air."[61] The very insulation of this world—its isolation from other worlds and possibly from the sun—is responsible for its decay. Supplies of energy are limited. And like Mr. Turveydrop, ladies and gentlemen of the fashionable world are bent on keeping down the world's realities, on suppressing the loss of distinction, the equilibration occurring spontaneously around them. They have "found out the perpetual stoppage," the steady state that like Spencer's *moving equilibrium* at once approximates and staves off the actual stop, the final equilibrium of heat death. What energies they do have must be directed to keeping "everything . . . languid and pretty."[62] They cannot do more than hope to prevent further decline.

At the center of this world are, of course, Sir Leicester and Lady Dedlock. If the Dedlocks are not quite dead, they certainly approximate equilibrium very nearly. Having achieved a stagnation trumped only by Miss Havisham herself, the Dedlocks have perfected the art of accomplishing no work. They have, it seems, only enough distinction to maintain the status quo, not enough to do useful work and not enough to garner usable energy. To Mr. Guppy, the clerk, their "family greatness seems to consist in their never having done anything to distinguish themselves, for seven hundred years." We have reason to believe him, too, since Mr. Guppy is rather sensitive to the energetics of personal interactions, guarding against drains on his own energy by choosing activities "of an unexciting nature, which will lay neither his physical nor his intellectual energies under too heavy contribution" such as gyrating on his stool and stabbing his desk with a pen knife. Nonetheless, he cannot avoid depletion and later laments "the energy [he] once possessed." And, like so many others, Guppy experiences the Dedlocks as a large energy sink. When he comes with a friend to have a look at their house, they find themselves, like so many others who tour the houses of the fashionable world, "dead beat before they have well begun."[63]

All but depleted, the Dedlocks deplete whoever comes into contact with them. Their many cousins might, except for their cousinship, "have done well enough in life," but their potentially purposeful energies are drawn into the energy sink as they "lounge in purposeless and listless paths." The central question that preoccupies the cousins, moreover, suggests the convergence of deportment and waste, for they are "quite as much at a loss how to dispose of themselves, as anybody else can be how to dispose of them."[64] Similarly, the energies of the Dedlocks' neighbor, Mr. Boythorn, are disposed of in his dealings with them. Mr. Boythorn's "unimaginable energy" as well as his "energetic gravity" are directed into anger, with both Chancery and Sir Leicester. He engages with the latter in an almost mechanical push-pull of trespass, assault, battery, and consequent legal actions, even going so far as to deploy a "fire-engine" against Sir Leicester's men.[65] But this double contact with the Dedlocks and the law drains even his considerable energies; he manages no more with all his efforts than to maintain the status quo. The Dedlocks thus draw on neighbors, relations, and incidental passersby with only this effect: they get to stay right where they are, forever distinguished as the Dedlocks of Chesney Wold. For Sir Leicester, this is the only work that matters.

The Source and the Sink

Unlike mechanical work, social work thus proves rather hard to pin down. *Bleak House* proves full of usable energy—all of it driving inevitably to disorder. It is full of engines as well, full of source/sink pairings that determine whether that energy is well-, mis-, or even peculiarly directed. Whether we consider the Jellybys and Africa, the Pardiggles and the poor, or another source/sink pairing—Lady Dedlock and the family lawyer, Mr. Tulkinghorn—the question has shifted not to whether disorder is produced, but whether work has been done in the process. The answer depends, of course, on whom we ask. Mr. Tulkinghorn's opinion is clear: his work is to maintain the Dedlocks in their current state of distinction. Whether this constitutes useful work is nonetheless under contention. His first conversation with Lady Dedlock indicates not only that they disagree about the usefulness of his work, but also that both are aware of this central question of energy use. When Lady Dedlock comments about her "cause" that "it would be useless to ask . . . whether anything has been done," Mr. Tulkinghorn answers in a typically evasive manner, "Nothing that YOU would call anything has been done to-day." Her "nor ever will be" seems correct, for her cause is none other than Jarndyce and Jarndyce. And it is perhaps a good thing for Sir Leicester that he "has no objection to an interminable Chancery suit,"[66] for it is all part of the perpetual stoppage.

It is at first, however, rather difficult to determine how Lady Dedlock functions in the draining environment of Chesney

Wold. Her cold facade, her not "melting, but rather freezing mood," her "exhausted composure, a worn-out placidity, an equanimity of fatigue"[67]—a veritable treasure chest of entropy words attach to Lady Dedlock!—seem to mark her as without useful energy. She appears to have succumbed to the draining effects of Chesney Wold or even to be a sink herself. But she is faking. She has done useful work and will do so again. Her fatigue masks energy sufficient to have produced Esther (yes, her daughter, though Esther does not know it) and to have kept that fact a secret, thus arguably driving the novel. When necessary, Lady Dedlock even finds the energy to free her maid Rosa from the drain of the Dedlocks. Meanwhile, she proves not so much cold as extremely well insulated, for most are unaware of the depths of her energy.

There are, of course, exceptions. Mr. Guppy—fatigued by contact with the law and the Dedlocks—recognizes it at first sight. When, during his tour of Chesney Wold, he is so low that "even the long drawing-room . . . cannot revive [his] spirits" and he "droops on the threshold [with] hardly strength of mind to enter," he warms immediately to the portrait of Lady Dedlock, hung significantly over the fireplace. The portrait "acts upon him like a charm. He recovers in a moment." Guppy, moreover, responds instantly to Lady Dedlock's likeness to Esther.[68] His attraction to Esther and his response to the portrait are of a piece: both energize Guppy, for both women are considerable sources of energy. As Guppy springs into action and discovers the secret of Esther's birth, he also initiates the novel's energetics of secrecy. For in discovering Esther's parentage, he also discovers Lady Dedlock's secret stores of energy and draws a bit of this to himself.

Mr. Tulkinghorn, however, will go far beyond Guppy in attempting to draw and direct Lady Dedlock's energies. In discovering the secret of her motherhood, he too uncovers her secret stores of energy. He is explicit about this discovery, thinking as he watches her struggle, "What power this woman has to keep these raging passions down!" and then "The power and force of this woman are astonishing!" For all that Lady Dedlock herself seems cold and drained, Tulkinghorn knows that the things that make her appear cold—her composure, her indifference—are energy in another form. Recognizing the extent of Lady Dedlock's energies as well as her efforts to insulate them, Tulkinghorn thinks this "woman has been putting no common constraint upon herself." It turns out that even Tulkinghorn does not know the full extent of her energy, though he might if he could see her "hurrying up and down for hours, without fatigue, without intermission."[69] By contrast, Mr. Tulkinghorn himself is closely associated with entropy. For seeing him in his office, "in his lowering magazine of dust," we are forcefully reminded of the decay of all things, "the universal article into which his papers and himself, and all his clients, and all things of earth, animate and inanimate, are resolving." Embracing the decay that surrounds him, "Mr. Tulkinghorn sits at one of the open windows enjoying a bottle of old port," itself a figure for decay like the dust that he "scatter[s], on occasion, in the eyes of the laity."[70] This increase of entropy cannot be divorced from the decrease of usable energy. If one particular form of disorder Tulkinghorn creates is scattered dust, the particular forms of energy he absorbs are not only the metaphorical energy of secrets, but also the literal energy of visible light: "One peculiarity of his black clothes, and of his black stockings, be they silk or worsted, is that they never shine. Mute, close, irresponsive to any glancing light, his dress is like himself." He absorbs more than light, however. He absorbs Lady Dedlock's energy. Even conversation with him draws from her stores, as she is "obliged to set her lips with all the energy she has." And if it seems, as their deadening interview draws to its end, as if the coldness of the stars beginning to pale froze her, appearances deceive once again. The immediate freezing effects come from conversation with Mr. Tulkinghorn, who "when the stars go out" looks very near his own final rest. From a broad narrative perspective, Mr. Tulkinghorn is a necessary evil. From the perspective of the novel-as-engine, he is the most efficient heat sink around. Like Watt, Dickens finds that his engine is most efficient when he introduces a contained cold spot, a condenser. Neither enemy nor friend, but "one who is too passionless to be either," he is, as Lady Dedlock observes, "mechanically faithful." And mechanically speaking, a lawyer is a sink: "His calling is the acquisition of secrets, and the holding possession of such power as they give him." In the novel's emerging energetics of secrecy, he has a remarkable ability to draw secrets and therefore power toward himself. And as a conscientious condenser, he is intent on doing work as he directs the dissipation of Lady Dedlock's energies. Her fall, if he has his way, will serve the preservation of Sir Leicester and the status quo at Chesney Wold.[71]

Together, then, Lady Dedlock and Mr. Tulkinghorn provide an elaborate prototype of the novelistic engine. Each enhances the primary function of the other; their pairing insists on the simultaneity of using energy, doing work, and increasing entropy. Tulkinghorn recognizes and draws out the energy within Lady Dedlock, showing her resistance to the fatigue he himself promotes. Lady Dedlock, after a long period of insulated stagnation, will do work. It is not, however, the work Tulkinghorn intends. For without wishing to, he accelerates her release of Rosa. In helping Rosa to marry Watt Rouncewell, Lady Dedlock does real work, increasing Rosa's potential by elevating her (energetically speaking) to a place of greater productivity. Of course, wherever work is done, in *Bleak House* and beyond, in the social as well as the physical universe, it is with the now familiar qualifiers: Lady Dedlock cannot do as much work as we would hope (she fails, at the end, to reunite with Esther) and always at the expense of greater disorder (leaving her dead and Sir Leicester much the worse for wear). But Lady Dedlock finally decides what limited work she will do; Tulkinghorn controls only the rate, not the product of her fall. This engine thus provides hope that the individual may find, in the apparently deterministic universe of ever-increasing entropy, a small space for useful work where she may direct her own energies.

A Thought on Machines: Fact and Fiction

A scientific fact is a statement, made in a certain way and accorded a certain reception. Rather than diffusing on its own, as they so often seem to do, scientific facts require quite a lot of help to achieve the status of fact and to be spread about, both within and outside the community we call scientific. Much of this help comes from nonhuman *actants*, whose responses to a variety of *trials* constitute the list of features that become the fact or the scientific object.[72] Much of the help, however, must come from people, who not only set up most of the trials, but also provide the language out of which all lists and all statements, factual or otherwise, are made. And the novel, abundantly stocked with language, proves at once a cog in the complex machine that makes and circulates facts as well as a machine in its own right.

To make sense of this grand metaphorical gesture, it will help to consider what we mean by a *machine*. "As its name implies," Bruno Latour tells us, a machine "is first of all a *machination*, a stratagem, a kind of cunning, where borrowed forces keep one another in check so that none can fly apart from the group."[73] Machines are closely associated with automation. And for a familiar example of an automaton, he cites none other than our old friend, the Newcomen steam engine. The various parts of this engine work to link, through a long series of associations, "the fate of coal mines to the weight of the atmosphere." The actors include a *piston*, drawn by a *partial vacuum*, produced by "*condensing steam*, pushed by *atmospheric pressure*, [which lends] its strength to the *pump* that extracted the *water*, that flooded the *coal mine*." It sounds like an engineer's drinking song. But the more we think of a machine as the complex strategies that link the fates of disparate elements, which may be "freely chosen among human or non-human actors," the more its similarity to a novel becomes apparent. And the more we understand how "the engineer's ability lies in multiplying the tricks that make each element interested in the working of the other,"[74] the more we think of Dickens.

Bleak House, then, is the machine[75] that links the fate of the dead man, who had been kind to the boy who sweeps the crossing, who provided information to Lady Dedlock, who turned out to be the mother of our heroine, Esther Summerson. Doubly so because the dead man is revealed to have been Captain Hawdon, the erstwhile lover of this same lady and Esther's father. *Bleak House*, it seems, is a particularly complex machine, with hundreds of elements intricately linked and placing checks on one another, and chock-full of simpler, smaller engines, as we have seen.[76] It is, moreover, not merely a tool—"a single element held *directly* in the hand of a man or a woman"[77]—of its author, but an apparently self-regulating mechanism that, like an engine, incorporates its own source and sink.

Faint echoes of the epistolary linger in *Bleak House*, because Esther's part of the narrative is penned to an anonymous correspondent, apparently at his or her request. Esther's modest interjections and occasional worries about her narrative difficulties, recall the early novelistic practice of writing in the form of letters. Nineteenth-century refinements of free indirect discourse or omniscient narration numb our senses, making it feel as if the narration comes from everywhere or nowhere and diffuses independently of our help. In this way, fiction may stage the apparently spontaneous diffusion of fact. But Esther's gestures toward "any one who may read" and her tantalizing final reference to "the unknown friend to whom [she] write[s]" remind us that someone must write and someone else must read. Thwarted in her first chapter and last by men who direct her toward a mirror, she proves unable to keep herself out of the narrative, to keep her "little body [in] the background."[78] And together, she and her unknown friend remind us that to work, a narrative engine must have both a source and a sink.[79]

6   Bodies in Heat: Demons, Women, and Emergent Order

Chaos is the womb of life, not its tomb.
—N. Katherine Hayles

What is life but organized energy?
—Arthur C. Clarke[1]

The late nineteenth century evinces a sense of decay so prevalent that we refer to the period as *decadent*. This general sense of grand-scale decay is reflected in texts that represent the earth as succumbing to heat death, such as H. G. Wells's *The Time Machine* and Camille Flammarion's *Omega: The Last Days of the World*. In turn, accounts of literature and history reflect this expectation of the late Victorian affect associated with energy physics.[2] With twentieth-century chaos and information theory, however, entropy emerges as a thing of potentially positive value. But perhaps this is less of a surprise than it at first may seem. Certainly, we understand *decadence* as evoking not only the problems, but also the problematic pleasures of dissipative behaviors. So too do late Victorian thermopoetics have a productive and pleasurable side. Kelvin himself imagined "everything in the material world [to be] progressive" before he adopted the language of dissipation.[3] And "it is a mistake to imagine the second law as having to do only with dissipation. In the interpretation of its founders in Britain, it was just as much about the conditions for the progress of civilization and the moral duty of man to maximize the utility of the productive forces available in nature."[4] Maximizing such forces, of course, had long driven improvements in engine design. And even the second law could not fully extinguish the hope of a perfect engine, which would produce work and increase order without producing waste.

Indeed, some wondered whether the second law was really all that unyielding. Perhaps entropic decay was not so much inevitable as really *really* likely. The question was addressed by James Clerk Maxwell, as he worked to help develop *statistical mechanics*. In this mathematically sophisticated version of thermodynamics, entropy became a statistical measure of the disorder within a system—a measure of the likelihood that the particles of a system would arrange themselves in a messy way (with hot/fast and cold/slow particles all mingled together) rather than an orderly way (with distinguishable hot spots and cold spots). Maxwell went so far as to imagine a creature, later dubbed a demon, who could sort hot particles from cold, thus decreasing the entropy of a system and suggesting that dissipation is not a universal law, but only a macroscopic phenomenon born of an overwhelming statistical tendency.

Maxwell's demon has caught the fancy of subsequent generations of scientists, literary scholars, programmers, and gamers.[5] "One of the most famous conundrums in the history of science,"[6] the demon possesses a capacity for infinitesimally minute observation and direction quite beyond human capacity. As such, it implies that our inability to circumvent the second law is a matter of human limitations; we're just too big—and not nearly observant enough. But the demon also drives changes of interpretations of entropy and indeed chaos itself, facilitating the move from a version of chaos that figures as the opposite of order to one that understands chaos as indicative of deep, emergent, dissipative, or otherwise alternative structure. In both ways, the figure of Maxwell's demon takes his place beside the entropic individuals of late-century fiction, who can control energy in ways others cannot, building even better engines. Demonic figures like Dorian Gray and Dracula effect the decrease of entropy, if only locally, for some alternative definition of what constitutes work. They do so, moreover, in ways that baffle mere human senses and make us wonder whether the second law of thermodynamics is so absolute after all.

The Entropic Individual

We have seen with Lady Dedlock and Mr. Tulkinghorn that there may easily be disagreements over the path a system should take as its energy is dissipated, over what work is. And the possibility of an individual particularly skilled at concentrating and using energy in a late-entropic state of affairs—what I like to call an *entropic individual*—may be fraught with anxiety as well as promise. After all, if we believe the second law of thermodynamics (and we generally do), an entropic individual can only concentrate usable energy and increase local order by creating disorder in the larger system. And of course, the entropic individual may, like Lady Dedlock, have her own conception of how to use the energy she has. Late Victorian fiction features quite a few such entropic individuals, among them Dr. Jekyll, Wells's Time Traveler, Dorian Gray, and Dracula.

A self-consciously entropic individual, Dr. Jekyll understands the creation of local order—his upright moral successful self—to be dependent on the increase of disorder elsewhere. We may understand Hyde as the increase in entropy necessary to the higher (energy) state enjoyed by the upright, vice-reduced Dr. Jekyll. While each transformation involves the doing of work and increase of order (in the sorting of good from evil), a bit more waste (recall Q_c) is produced each time, accounting for the growth of the entropic Hyde. Allen MacDuffie has written compellingly on the thermodynamics of *Dr. Jekyll and Mr. Hyde*. He focuses on the impossibility of sustained reversibility—that aspect of the second law that suggests that in any process involving the transformation of energy (which is to say, any process), some energy must be lost to unusable forms. It is for this reason that the second law confirms what we already knew: there is no perpetual motion machine. Understanding the Jekyll/Hyde pairing as an engine, MacDuffie suggests that the second law prohibits the indefinite continuance of these transformations.[7] Nonetheless, the fact that Jekyll does it may be even more impressive than the fact that he cannot do it forever. For as long as the engine is operating, Hyde acts as a very efficient condenser. Like Tulkinghorn, he serves as a repository of the disorderly, drawing it away from Jekyll before releasing it into the larger system.

Wells is all over this. Even Dr. Moreau, who is less astute about energy than anatomy, nonetheless suggests how much work must be done to increase complexity of organization, a form of ordering. The tendency of his beast-people, moreover, to spontaneously revert to indistinct, homogeneous, or otherwise mishmashed forms (with the concomitant decrease in social order) posits indifferentiation as a more entropic state. This reversion has the ring of Spencerian dissolution; it is a descent from an orderly heterogeneity to a disorderly homogeneity. But devolution presents itself most explicitly as enervation in *The Time Machine*. The Time Traveller's final glimpses of the future, the views of the "dying sea" and "darkling heavens," of "the sun grow[ing] larger and duller in the westward sky, and the life of the old earth ebb[ing] away," epitomize Victorian depictions of the thermodynamically ordained end of the world, the final surrender of all things to cold, darkness, and silence.[8] But Wells complicates his depiction of entropy as time's arrow by reversing time itself. After some maximum period of local order, later dates may indeed be recognized by the level of degeneration apparent in the measure of energy beyond our use. But as with all prior and succeeding generations of time travelers—and in the event, who can tell which is which?—Wells's Time Traveller suggests that the notion of *later dates* is not itself irreversible. If one can reverse the direction of time, one can reverse the direction of entropy. And the Time Traveller returns. That he eventually disappears to return no more suggests something about his counterentropic gestures that echoes those of Dr. Jekyll: such reversals cannot be sustained indefinitely. But there is nonetheless something intriguing in the fantasy that they may occur at all.

Entropy: This Time It's Personal

In *The Picture of Dorian Gray* (1890), entropy is fully integrated with the discourse on aging. Though expressly interested in science, especially chemistry, *Dorian Gray* rarely makes explicit use of the language of energy. It is, after all, unnecessary. It is already a *fact*. The action of entropy is apparent in the action of time on all systems—personal, historical, and material. In the decay of cultures, the passing of seasons, the wearing thin of tapestries, the aging of individuals, and the dissolution of all material things, entropy serves throughout *The Picture of Dorian Gray* as time's arrow. Lord Henry, moreover, makes it personal. "Time is jealous of you," he tells Dorian, "and wars against your lilies and your roses. You will become sallow, and hollow-cheeked, and dull-eyed. You will suffer horribly."[9] He is, of course, merely applying the general rule to the individual case. In the passage of time, entropy must increase, and we all "degenerate into hideous puppets." Why should Dorian be different?

But Dorian is different. Everything decays around him, but he escapes. "Almost saddened by the reflection of the ruin that Time brought on beautiful and wonderful things," Dorian speculates on his own relative success in resisting entropic decay: "How different it was with material things! Where had they passed to?" Though Dorian is not clear on how, the picture does indeed enable him to stave off the increase of entropy in his person. As he looks at the picture "with a feeling of almost scientific interest," he speculates that there may be some "affinity between the chemical atoms" and his soul, that somehow they realized his dreams. "Or," he worries, "was there some other, more terrible reason?"[10]

Indeed, there is a far *less* terrible reason. For the picture acts as a heat sink. Certainly, the entropy of the system must increase. But as we have seen, what is true of the whole may not be true of the parts. Order may be preserved—even increased—in one part of the system as long as entropy increases overall. Dorian and his picture then figure as coupled parts of a larger system. Like Tulkinghorn and the condenser, the picture serves as a localized site for the rapid increase of entropy. Subject to a process of decay that is more rapid even than ordinary aging, the picture enables both the preservation of Dorian in a youthful state and whatever constitutes Dorian's work. Whether this latter appears in the ordering processes of collecting, sorting, and resorting—as in the "settling and resettling in their cases the various stones he collected" to which Dorian devotes whole days—or in the grander object of developing a "new Hedonism that was to recreate life,"[11] Dorian's work is about creating and maintaining order in resistance to the ordinary action of time.

A Brief Chaotic Interlude

At the same time, Dorian suggests possibilities of transformation that seem to defy the second law through forms that eventually find their way into science in a field known as chaos theory. Waking from the "unreal shadows of the night," Dorian embodies the choice wherein we experience a "terrible sense of the necessity for the continuance of energy," which may manifest itself "in the same wearisome round of stereotyped habits" or in the creation of strange new worlds.[12] The former evokes a version of time and expenditure of energy that in the Victorian moment is gendered decidedly circular, feminine, foreign, regressive, and entropic. The latter, however, is not the corresponding, masculine imposition of linear order generally taken as its binary opposite.[13] In a counterentropic vision (consistent with other modes of gender bending in *Dorian Gray*), Dorian imagines a new possibility for the continuance of energy. He fantasizes that "our eyelids might open some morning upon a world that had been re-fashioned anew for our pleasure in the darkness, a world in which things would have fresh shapes and colors, and be changed, or have other secrets."[14]

For a reader passingly familiar with chaos theory, Dorian's fantasy of creation is uncannily suggestive. Chaos theory evolved in two branches, each of which articulates a surprising relation between chaos and order: "In the first, chaos is seen as order's precursor and partner, rather than as its opposite. . . . The second branch emphasizes the hidden order that exists *within* chaotic systems." Dorian's notion that new worlds may be created from the darkness is a counterentropic development that evokes the first kind, the "something-out-of-nothing" branch of chaos, which focuses on the "spontaneous emergence of self-organization from chaos." Out of the darkness arise what we might call—in parlance that evokes both this kind of chaos and decadence—*dissipative structures*. Or perhaps those structures were already there in the darkness. That things in Dorian's new world "have other secrets" seems to render them akin to the second branch, in which chaos is distinguished "from true randomness, because it can be shown to contain deeply encoded structures called 'strange attractors.'" Unquestionably a strange attractor himself, Dorian evokes the near impossibility of prediction within the nonlinear systems that are the focus of this second kind of chaos, for in his new world "the past would have little or no place." We needn't worry much, however, about the distinctions between the two branches of chaos theory. Neither would be articulated for over a half century, and Dorian's fantasy does not clearly distinguish between them.[15] But the fantasy itself is important, and as we will see, Dorian's is not the only fantasy of a new kind of order, a new valuation of work, the creation of order from disorder, pleasure in darkness.

Even for Dorian Gray, however, it is only a fantasy. In spite of the delicious echoes of decadence in general, *Dorian Gray* in particular, in such terms as *dissipative structures* and *strange attractors, Dorian Gray* is not chaos theory. However much Dorian may wish to create worlds out of darkness, he cannot in fact do so. Though Dorian may be said to have made great strides in engine design, he finds himself still bound by the limitations of traditional thermodynamics. The increase (even the maintenance) of order in Dorian's person, his collections, his fetishizing of the beautiful, can take place only very locally and only at the cost of increasing entropy elsewhere in the system. The rapid decay of the picture as localized heat sink strongly evokes the function of the narrative condenser. But *Dorian Gray* also evokes the engine in its broader effects, for all engines heat up the world. Dorian and his picture are not perfectly insulated, and Dorian's "creation" of usable energy and order comes at the expense of disorder in the greater system. Thus, the harm that is attributed to Dorian's "influence" undoubtedly includes but is not limited to the homophobic anxiety of contamination. Energetically speaking, this harm is the consequence of the local increase or maintenance of order. If the decay of the picture is reversible in the sense that it returns to an orderly state, Dorian's aging exactly compensates. But usable energy must be lost in such reversals, and the balance between the picture's restoration and Dorian's aging is belied by the net increase in entropy effected throughout the novel. The corruption of everyone with whom Dorian becomes intimate is indicative of widespread decay. This decay can be seen in the dissipation of men like Sir Henry Ashton, Adrian Singleton, and the young Duke of Perth; the loss of reputation of women like Lady Gwendolen; the deaths of Sybil and James Vane, Hetty, and Alan Campbell; and the literal dissolution of Basil Hallward. The production of disorder in the novel, moreover, is echoed by the production of disorder in the larger system within which the novel is produced, for *The Picture of Dorian Gray* itself nucleates a scandal. Considered a dangerous book for its effects on young men, it was used against Wilde during his trial for "gross indecency," after which Wilde was thrown into prison for sodomy.[16] But one might say that he was guilty of doing what, in *Bleak House*, Skimpole imagined of Jo, for (mis)directing his energies into the production both of poetry and of disorder in the "general arrangement of the whole system of spoons." Unlike Jo, Wilde would not have hesitated to seize a spoon.

Vast Engines of Enlargement

Like *Dorian Gray*, *Dracula* emerges from a sense of widespread decline. Both suggest that there is not that much energy— sexual, political, usable—to go around. Fraught with discomfort that is at once sexual and imperial, *Dracula* is riddled with women who become "suddenly sexual" against the backdrop of an empire in decline and under siege. The text is cognizant, moreover, that not only sex and empire, but also energy and technology are at issue.[17] The technologies at issue, however, are different from those that drive *Dorian Gray*, for unlike Dorian, who seems to operate in a closed system, which he most effectively heats up, Dracula is from the outside. A "coming race" unto himself, he represents at once the colonized and the colonizer, the potential dangers of the British imperial project and the fear that someone else may best the British in building imperial engines.

The novel certainly conceives of empire building as a feat of engineering, though it carefully mediates this belief through the mouth of a madman. The United States, it seems, may have hit on the better technology, for "the power of Treaty may yet prove a vast engine of enlargement," which will fully realize the Monroe Doctrine, such that "the Pole and the Tropics may hold alliance to the Stars and Stripes."[18] Vampirism functions more tacitly as technology—both reproductive and imperial[19]—all the more horrifying for not announcing itself as such. In stark contrast to the difficulties faced by the Westerners who oppose him, Dracula accomplishes a remarkable amount of work. He is, after all, producing vampires at an alarming pace—through a process with implications at once imperial, reproductive, sexual, and energetic. Were his detractors possessed of a similar capacity to control the flow of energy and do work, they would undoubtedly deem it useful. In Dracula's hands, vampiric technology reveals its double-sided demonic potential—accomplishing what we would wish to do, but with a scandalous circumvention of (the second) law.

By contrast, the vampire hunters seem wasteful, enervated, and extremely slow to reproduce. But they are not without energy. It seems rather that they don't know how to use it. Their mistake, as Dracula points out, is that they spread these energies too thinly; "they should have kept their energies for use closer to home."[20] This ambiguous line suggests not only the misuse of masculine energies in the failure to protect the one woman among them, Mina Harker, but also the inefficiency of too-widespread imperial endeavors. Energies that are too widely disseminated—sexually or imperially—run out too fast. Indeed, the second law suggests that empire can't last: "In Kelvin's prose, the rhetoric of imperialism confronts the inevitability of failure."[21] On the one hand, *Dracula* manifests the belief, also held by the Vril-ya, that opening the system, aka expanding the empire, is necessary to sustaining a moving equilibrium within, necessary to the maintenance of a culture. On the other hand, the text recalls Mrs. Jellyby and the natives of Borrioboola-Gha, enacting the fear that the empire will constitute an enormous heat sink that drains the energies of the culture it was intended to sustain. *Dracula* enacts a precarious balance between openness and isolation.[22]

It seems, then, that one must have access to an outside, but opening the system can be a dangerous business. From the perspective of expanding empire and superstition alike, the vampire must be invited in. And so a nice English solicitor by the name of Jonathan Harker travels east to Transylvania to help Dracula relocate, ensuring that "the empire fans out across the globe, collecting its grab-bag of completely incongruous possessions, while at the same time the maintenance of a national community back in the metropole, as it were, siphons off tremendous amounts of ideological energy."[23] In the comfortable assurance of Western superiority, Harker heads east as both tourist and businessman, collecting folklore and recipes, and seeking an influx of Eastern wealth that will help pay his salary as well as revive a decaying English property. His comfort, however, is short lived. For he finds instead that Dracula wants to draw from him: information, energy, blood. Dracula himself thus functions as a localized heat sink, who specifies, reiterates, and potentially reverses the various energetic functions of empire. Not content to stay nicely in his place in the East, the proper object of the touring Westerner or, at best, the source of information regarding the picturesque objects of Western curiosity, he wishes to occupy the metropolis himself. While his energies are necessary to keep the imperial engine running, he nonetheless threatens to mess with the direction in which energy flows.

Condensers Within

In the paper that announces the "Universal Tendency in Nature to the Dissipation of Mechanical Energy," William Thomson observes that "there is an absolute waste of energy *available to man* when heat is allowed to pass from one body to another." He further observes that unless something changes radically, the earth must at some time become "unfit for the habitation of *man as at present constituted.*"[24] I have shifted the usual emphases of these statements, from the focus on the operation of universal laws, to their specific impact on humans as we know them. And while the former emphasis may explain the wish to discover hitherto unknown sources of energy (among them radiation),[25] the latter suggests that what will change—for better or worse—is not the universe, but man. It then follows that late-entropic fictions should emerge that, like the story of the demon, imagine beings similar to humans, but constituted sufficiently distinctly as to be able to better manipulate the decreasingly available energy of the environment. The cognizance of universal decay thus fosters the proliferation of demonic individuals whose entropy-reducing ability is at once desired and feared.

Combining the condensing function of Mr. Tulkinghorn with the imperial intentions of the Vril-ya, Dracula proves just such a demon. His very ancientness has schooled him in the efficient use of energy apparently almost past its usefulness. Repeatedly associated with the setting sun, he evokes not only the end of the British empire, but also the ability to move once the obvious (point) sources of energy are withdrawn. The vampire's better ability to move after dark—"for at sundown the Un-Dead can move"—suggests how efficiently he draws energy from a failing system.[26] And Dracula's triumph at seeing the "sinking sun" proclaims his confidence in his own thermodynamic proficiency. If he cannot properly be said to live, his un-dead state recalls the Vril-ya's closeness to final rest. Dracula's death is represented as a "moment of final dissolution" with which he is at peace. But his final moment bespeaks not only his salvation, but also the realization of complete entropy, as his "whole body crumbl[es] into dust and pass[es] from our sight."[27] Dracula's entrance to the Unseen Universe is a transformation of energy that emphasizes his close association with entropy. Not only has he wrought the obvious disorder associated with his violent past, but he also makes it impossible for potential victims and readers alike to maintain a variety of operative distinctions: life/death, male/female, blood/milk/semen, self/other, and so on. In the interim, Dracula finds energy in a failing system and attempts to direct it.

In the failing system of *Dracula*, energy may yet be found in a variety of forms: Van Helsing's "usual recuperative energy" and Renfield's "mental energy," as well as the earnest and energetic efforts of the vampire hunters in general. Harker's energy is here "volcanic" and there "desperate," but mostly, we are simply assured that he is "full of" it.[28] In fact, no character is described in energetic terms as often as Jonathan Harker— something of a surprise since we also witness him ill, distraught, and once in an outright stupor. Perhaps it is this alternation, wherein the text marvels at his energy and stages its loss, which gives his descriptions something of a protests-too-much feel. But this mild-mannered solicitor is "full of energy and talent in his own way" and "uncommonly clever, if one can judge from his face, and full of energy."[29] In sum and individually, Harker's various forms of energy suggest its transformability—sometimes manifesting itself as discretion and silence, other times as resolution and strength. Its ebb and flow further establishes this transformability. On the trail of Dracula, he finds that sleep transforms him. "I am a new man this morning," he observes; "I can feel that my strength and energy are coming back to me." This transformation reverses an earlier one, in which Harker's energy seems to be—but, in fine first-law fashion, is not—lost. One night, he is "a frank, happy-looking man, with [a] strong, youthful face, [and] full of energy"; the next day, he is "a drawn, haggard old man."[30] Harker's rapid aging correlates with his apparent loss of usable energy, or what is the same thing, increase of entropy. The flow of energy here is in the usual direction. Entropy, time's arrow, is flying in the right direction, if faster than usual. We know, moreover, that it is Dracula who accelerates and directs Jonathan's loss of usable energy, elsewhere actually placing him in a "stupor such as we know the Vampire can produce" and, as it were, leaving him cold.[31] And indeed, Dracula draws the power of others to himself as well, diminishing the energy (especially the thermal energy) of the environment. Thus, as Dracula comes, Mina finds herself "powerless to act," finds the air itself "heavy, and dank, and cold" and everything dim around her, the "gaslight . . . only like a tiny red spark through the fog."[32] And though, as we will see, Dracula does some rather surprising things with the energy he thus draws to himself, he seems at first to operate simply and thermodynamically as a heat sink, bound by ordinary rules of conservation and dissipation.

Meanwhile, Jonathan's aging bespeaks a conservation principle with positive as well as negative consequences. For we are assured that "his energy is still intact," not actually gone, but merely transformed. "In fact he is like a living flame,"[33] a source of heat energy, which on the one hand is so often the byproduct of apparent energy loss, but which, on the other hand, can be at least partially recovered, as by the employment of any heat-driven engine. And indeed, when Harker finally recuperates his energy in the final chase scene, he immediately turns to "minding the engine." As a new man, he shows himself once more capable of producing work, but also expresses concern that the engines (his own and that of the steam launch) will prove insufficient, incapable of transforming heat into the work that needs doing: "If we could only go faster! but we cannot; the engines are throbbing and doing their utmost." In the final stage of the hunt for Dracula, we witness the struggle of heat against cold, the furnace that drives the steam launch and warms its crew, the cold of the river that threatens to absorb the energies of both. Very exciting stuff. Even "Lord Godalming is firing up" in an ambiguous way that suggests that the working of the steam engine is tied to the recuperation of his energies, as well as Harker's.[34] This local reversal of entropy, the gathering of heat energies in the production of useful work, is reiterated by the novel's final product: the Harkers' son Quincey, the only child of the only marriage that survives *Dracula*. At considerable cost, the vampire hunters have managed to harness usable energy, to reverse the increase of entropy (however locally), even (in this specific energetic sense) to reverse time and restore order.

It's a big job, but Dracula, for one, has already done it and with much less apparent effort. His demonic ability to decrease entropy becomes manifest when he shows up in England as a new man, or at least "grown young"[35]—having effected in himself the opposite transformation that he triggers in Jonathan. As if linked by a conservation principle, the attacks that age or exhaust his victims, strengthen Dracula. The energy thus drained takes a variety of forms, sometimes invigorating his material body, sometimes concentrating itself in the mist that is Dracula, a concentrated "sort of pillar of cloud" that emerges from the smoke that pours into Mina's room "with the white energy of boiling water" while a "leaden lethargy" binds her will as well as her limbs. Dracula proceeds to demonstrate that he is able to concentrate diffuse energy into a usable form, gathering and directing energy, dividing fire, and ultimately transforming energy into matter itself, a "white face bending over me out of the mist."[36]

Of course, the most basic sensitivity to late Victorian thermodynamics suggests that blood functions as a figure for energy in *Dracula*. Lucy Westenra, for example, is repeatedly revived by its infusions, enervated by its loss. And certainly energy works very much like blood. It has, as one of my colleagues puts it, the quality of *stuff*. It is measurable. It flows. It is capable of transfer. One can treat it, both conceptually and mathematically, as fluid.[37] If we follow the trail of blood, tracing it momentarily as the fluid-flow analogy to energy, we find that Dracula draws it together and to himself: "Your girls that you all love are mine already; and through them you and others shall yet be mine."[38] To get the most out of a system far advanced into disorder, Dracula channels energy into a localized source, drawing it all together first through Lucy and, he anticipates, through Mina. In each case, the woman functions not only as a source of energy, but also as the small hole Dracula opens and closes to control the flow of energy in the form of blood.

The Sorting Demon

If Dracula's methods of garnering and using energy are not obviously possible, they are nonetheless compelling. The more we imagine Dracula as a better engine, the more suggestive he becomes of one Victorian fantasy of the perfect engine—the creature called Maxwell's demon. What the demon does is conceptually simple: it sorts. The creature, later dubbed a demon by William Thomson, is introduced in 1867 by James Clerk Maxwell to his friend P. G. Tait and described in a note to his *Theory of Heat* (1871):

If we conceive a being whose faculties are so sharpened that he can follow every molecule in its course, such a being, whose attributes are still as essentially finite as our own, would be able to do what is impossible for us. . . . . . . Now let us imagine [that a vessel full of air] is divided into two portions, *A* and *B*, by a division in which there is a small hole, and that a being, who can see the individual molecules, opens and closes the hole, so as to allow only the swifter molecules to pass from *A* to *B*, and only the slower ones to pass from *B* to *A*. He will thus, without expenditure of work, raise the temperature of *B* and lower that of *A*, in contradiction to the second law of thermodynamics.[39]

This being has a lot in common with the vampire: sharpened faculties, finite attributes, super- (or at least non-) human abilities. Moreover the demon produces a temperature difference—a higher-energy, lower-entropy state of the system. This "gradient making imp,"[40] moreover, produces without *expenditure*. That is, the demon accomplishes what usually requires work—the demon orders the system—without putting energy into the system.

Like the demon, Dracula seems able to produce work without expenditure,[41] to reproduce without the sacrifice of his own bodily fluids. He also takes it on himself to determine which boundaries will be crossed and how. A veritable Victorian triple point,[42] he resides on the "borders of three states" in an area of Europe so wild and unknown that it hasn't been mapped yet. Out of this disorderly environment, Dracula nonetheless proves well able to produce order. Even the road to his castle is generally a good one "different from the general run of roads in the Carpathians, for it is an old tradition that they are not to be kept in too good order." Similarly, his home decor reflects his ordering abilities; all of its fabrics are "centuries old, though in excellent order." Similar furnishings back in London at Hampton Court are "worn and frayed and moth-eaten"—a contrast that highlights Dracula's unique ability to resist entropic decay.[43] His most emphatic apparent exemption from the second law, however, shows in his capacity to evade death and to multiply the effects of that evasion as he produces more vampires. For all of the vampires remaining at the Castle Dracula—"those awful women growing into reality through the whirling mist in the moonlight"[44]—have shown themselves similarly capable of the energy-to-mass transformation we witness in Dracula himself. We assume, moreover, that each may also effect the transformations that preserve youth and produce new vampires. The terrifying efficiency of vampiric reproduction bespeaks the energetic efficiency of the mechanism itself. Dracula thus embodies at once the fear and the fantasy of a more efficient machine. With usable energy in short supply, he shows himself capable of gathering it, transforming it, and doing work with it.

Or what a vampire would call work. Vampiric reproduction gestures toward a valuation of the disorderly through which order emerges—or alternatively, in which we find a different kind of order. The "father or furtherer of a new order of beings,"[45] Dracula proves more of a strange attractor than even the beautiful Dorian Gray, not only because he inspires a desire so fully integrated with fear, but also because he shares the "odd combination of randomness and order that conveys the flavor of the strange attractor"[46]—that hard-to-describe set of points around

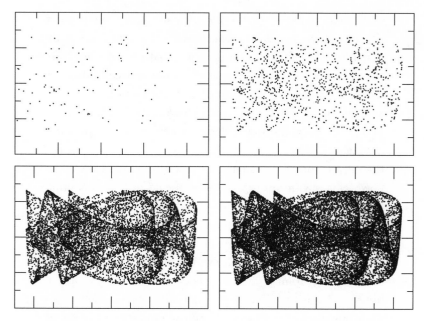

Figure 6.1
The Strange Attractor. As chaotic systems evolve, their behavior may seem random, but as we continue to track them, they eventually reveal a kind of nonrepeating pattern.

which a chaotic system swirls as it evolves (see figure 6.1). Dracula is indeed at the center of such a swirling, chaotic system—the "horseshoe of the Carpathians" around which "every known superstition of the world is gathered . . . as if it were the centre of some sort of imaginative whirlpool."[47] And Jonathan Harker hopes that the count will be the one to make sense of it all—a role with which he only partially complies. Indeed, as the father of "a new order of beings, whose road must lead through Death, not Life," Dracula further exhibits characteristics in common with the second kind of chaos, the some-thing-out-of-nothing version. Here "the essential change is *to see chaos as that which makes order possible*. Life arises not in spite but because of dissipative processes that are rich in entropy production. Chaos is the womb of life, not its tomb."[48]

Sexual Energies

And who better than Dracula to blur the distinction between wombs and tombs? His attraction is strange, because he draws to himself energies that others can't or won't use, in order to produce what they would not call work. That *Dracula* stages a contest for the control of women's bodies is well known. Our more specific interest here is that this contest is also about making use of available energy in a climate of scarcity. Dracula evinces a willingness to make use of women's energies at a moment when others are too afraid of it to recognize its use-fulness. In this way, the novel takes part in a profoundly am-bivalent discourse on women's energies. For while the laws of conservation and dissipation provided scientistic support for traditional views of women's roles at a time when these were changing rapidly, the new physics attracted more and more women into its fold. But where some resisted, other physicists welcomed, or at any rate allowed, women into their classrooms and labs.[49]

Similarly, the vampire hunters aren't sure what to do with Mina Harker. She is invaluable to them as the recorder, collator, and even translator of information. With her typewriter and her knowledge of shorthand as well as her journalistic recollection, it is she who orders all of the information that makes it possible to fight the vampire. (More on this role later.) At the same time, she is repeatedly excluded from the business of vampire hunting. As her husband puts it, "I am so glad that she consented to hold back and let us men do the work. Somehow, it was a dread to me that she was in this fearful business at all; but now that her work is done . . . she can henceforth leave the rest to us."[50] Jonathan's attempt to distinguish "her work" from "the work" suggests that the question of what work is, is still on the table and in ways that echo familiar Victorian genderings of work. Paradoxically, the kind of work that women generally did, including bearing and rearing children, was frequently deemed "unproductive labor," because it was considered to produce no salable product.[51] But Mina cannot do that kind of work if she is still chasing vampires; it's a consequence of conservation. So that when Dr. Van Helsing articulates the many reasons why "it is no part for a woman," he gestures toward the parts he finds most womanly—her heart, her nerves, and her womb—pointing out that because she is young and newly married "there may be other things to think of some time, if not now."[52] This polite allusion to future children rests on (indeed, works to factify) the belief that specifically feminine energies are subject to a conservation law. He thus mobilizes a rationale familiar to those medical men who, in working to maintain women in their traditional social place, drew on the conservation of energy for their rationale. After all, there was only so much energy to go around. Already sadly depleted by the demands of menstruation and the maintenance of female secondary sex characteristics, women, they argued, should preserve their remaining energies for the rigors of childbirth. To divert such meager resources into mental activity—for advanced education, the doing of physics, or the pursuit of vampires—threatened nothing short of the health of the nation.[53]

Patterns of repeated inclusions and exclusions suggest considerable ambivalence surrounding the question of what constitutes work. By excluding Mina, on the one hand, they seem to wish to distinguish what constitutes real work from what women may do. On the other hand, her periodic reinclusions reveal the men's anxiety that in a climate of scarcity, excluding her may well prove their undoing—that in leaving "the rest to [them]," Mina may well leave them to nothing but rest. Without Mina's energies, they might find themselves that much more quickly at a final state of complete rest, when no more usable energy is available to them. Thus, the acknowledgment that Mina has energy goes hand in hand with the desire to direct it. Even the attribution of a "man's brain" to go along with Mina's "woman's heart" seems to express the novelistic hope that her energies will be controlled in a masculine way, top-down and linearly, rather than running amok in decidedly female and vampiric ways.

When Mina does finally become a mother, the novel further reiterates its sense that human reproduction is bound by conservation laws. Once Mina's energies are devoted to the production of little Quincey, the novel makes it clear that she can't be both a mother and a vampire hunter. One role not only precludes the other, but also makes it "almost impossible to believe." We feel certain that though her son may someday know "what a brave and gallant woman his mother is," his understanding will center instead on "her sweetness and loving care" and the "men [who] so loved her, that they did dare much for her sake."[54] This last is a strange characterization of the events that lead to Dracula's demise, emphasizing the men's activity at the expense of Mina's by relegating her to motivation rather than actor in the story. Her work is no longer work.

Reordering: Chaos, Women, and the Dream of Information

According to Katherine Hayles, twentieth-century chaos theory dealt with a similar ambivalence in a similarly Victorian fashion. Noting that James Gleick's substantial and popular history of one branch of chaos theory is marked by a conspicuous absence of women, Hayles observes that "paradoxically this exclusion facilitates the incorporation of the feminine principle of chaos. ... admitting the feminine as an abstract principle but excluding actual women." This exclusion, which serves to keep "science . . . as monolithically masculine as ever," has a decidedly Victorian ring.[55] We are reminded of persistent attempts in *Dracula* to exclude Mina Harker while nonetheless making use of her energies. But as quickly as Mina is excluded from "the work," she is reincluded for the "special power" that the count has given her. And when in the final chase Van Helsing himself takes her into the heart of things, he makes it clear that it is because energy is in short supply. Relegating Jonathan to service on the steam launch because he is "young and brave and can fight, and all energies may be needed," he himself wishes to use "Madam Mina's hypnotic power" so that they may find their way, "all dark and unknown otherwise."[56] Thus her capacity to access and convey information once again requires her qualified reinclusion. The men find they cannot, in fact, do without her.

This is obviously true of human reproduction—though there are plenty of late Victorian fantasies that express the wish to circumvent the woman's role in this process as well.[57] But this is also true of other gathering and sorting processes—the ordering of information. For her "special power" is less a gift of the count than a new twist to a power she already has in spades, a gift, moreover, closely associated with her energy: "It is due to her energy and brains and foresight that the whole story is put together in such a way that every point tells."[58] Mina translates Jonathan's diary from the shorthand, preserves her correspondence with Lucy Westenra, transcribes Dr. Seward's phonographic notes, pastes newspaper clippings into her diary, and more or less collects every piece of written or spoken information that passes her way: "Just as all male blood passes through Lucy's veins, all their data passes through Mina's typewriter."[59] It generally does so, moreover, in triplicate, since Mina types it all up using what she calls "manifold" and (with a little help from Jonathan) puts it all "in order." No wonder such efforts draw the grateful exclamation of how "earnestly and energetically" they all have been working.[60] But the work of producing order has been almost entirely Madam Mina's.

Mina's capacity to access and convey information is closely connected to how her reproductive function is construed—paradoxically, in the same moment that seems to render one function incompatible with the other. For the cooperative paternity implicit in the child's "bundle of names [that] links all our little band of men together" suggests that Mina has functioned as a conduit for the pooling of masculine energies as well as for information.[61] The vampire hunters have learned something about compensating for a dissipated state of affairs by consolidating what usable energies remain into one place, a child who goes by the name of Quincey. If rather little and very late, it is nonetheless something. Thus Mina seems to have some of the capacities of Maxwell's sorting demon. But the "*bundle* of names" points to Mina's other capacity for consolidation, for hitherto in *Dracula* we have seen bundles of papers, letters, title deeds, and banknotes. Mina's demonic capacity to consolidate energy thus converges with her capacity to consolidate information—even more so as she approaches a vampiric state: "Here we step into the age of mechanical reproduction with a vengeance, since the reproductive process that makes vampires is so closely allied to the mechanical replication of culture."[62] The process by which Mina is (almost) produced as vampire and through which she will become able to produce more such, develops alongside and even reinforces her ability to convey and order information.

This second kind of sorting evokes a different kind of demon, Stanislaw Lem's "Demon of the Second Kind." This demon

is an up-to-date version appropriate to an information age. Like his predecessor, the Second Demon also presides over a box of stale air. Instead of sorting the molecules, however, he watches their endless dance. Whenever the molecules form words that make sense, he writes them down with a tiny diamond-tipped pen on a paper tape. Whereas the First Demon uses randomness to produce work, the Second Demon uses it to produce information.[63]

If Dracula himself is a demon of the first kind, Mina seems to vacillate between the first and the second—doing (re)productive work (potentially vampiric or human) and producing information, uncoding the narrative by her act of typewriting,[64] using her heightened sensitivity to identify sense and write it down. If Dracula, a traditional entropy-reducing demon, a hyperefficient machine on the cutting edge of science, evokes at once the possibilities of physics and its current limitations, Mina anticipates its future, the unimaginable possibilities of a science willing to admit both the female and the feminine, Marie Curie, chaos, and beyond.

Meanwhile, the second kind of demon suggests a relationship between entropy and information—a relationship that remains ambiguous and unsettled to this day.[65] Even the word *sort* is suggestive of this ambiguity, for while on the one hand it suggests the process of organizing, classifying, and ordering, on the other it is used to signal our inability to do so adequately—a *sort of* indicator that we aren't really sure. For us, a simple sort of explanation will serve to sort it out: If we count up all of the possible arrangements of the molecules within the "box of stale air," we may take that number (or more precisely, the logarithm of that number) as a measure of either the entropy or the information capacity of the system within the box. More arrangements or states means higher entropy. It also means more possible messages for someone like Mina to write down. It makes sense then that this late Victorian moment, so thoroughly steeped in the prevailing sense of entropic decay, is also a moment of increasing information, of more messages taking more forms. Little surprise then that we call on the second kind of demon to translate, direct, and order these messages, inviting more and more women into the public sphere as secretaries, stenographers, telegraph operators, and so on.[66]

Unless, of course, they are first invited to become vampires. The text, in fact, is obsessed with ordering information as the best hope for combating this entropically savvy demon. This obsession with information functions as an attempt to compensate for limitations of energy: "At stake is the hope that information will constitute a new regime, an economy of exchange not bound by the laws of conservation formulated for energy in the nineteenth century."[67] As Professor Van Helsing revs up the vampire hunters for their task (with Mina "act[ing] as a secretary"), he taps into this dream of information. He takes stock first of what the vampire can *do*, and then contrasts this primarily with what the hunters *know*. He thus attempts to place the vampire on the side of work and energy, the side of conservation and limitations, as "he who is not of nature [but] has yet to obey some of nature's laws." His pursuers, who can do a whole lot less, can nonetheless know a whole lot more. They have behind them "the forces of the information industry."[68] Indeed, even enumerating the vampire's abilities contributes to their stock of "data," transforming his entropic powers into information. It is clear, moreover, that they do so in order to compensate for scarcity in an energetic system, noting that the vampire's power to see in the dark is "no small power . . . in a world which is one half shut from the light."[69] That he can *see*, however, also evokes Dracula's attempts to know and to control the flow of information. So he regulates Jonathan Harker's correspondence from the Castle Dracula, works through several solicitors so that no one should know all of his affairs, and eventually burns not only the manuscript Mina had worked so hard to assemble, but also Dr. Seward's phonograph cylinders for good measure. This last move triggers Mina's reinclusion in the group's work, for as Van Helsing says, "It is need that we know all." So, after "ordering her thoughts," Mina tells them of her own vamping, and the race for information is on.[70] Dracula simply cannot compete. His letters can't compare with their telegraph; he does not know about "manifold" copies. The vampire's body may be "made to move in ways that anticipate the performance of bodiless communication technology," but Dracula himself does not have access to that technology.[71] Even as "the dream of freely flowing information," like the vampire, "strains to escape scarcity, restricted physical space, class, gender, embodiment, time and mortality,"[72] Dracula himself is excluded from that dream.

The Expansion of Fact

The vampire hunters, on their side, have the "power of combination, a power denied to the vampire kind." They succeed by mustering allies, many of them nonhuman. To an already mixed bag of alliances—including an English lord, solicitor, and doctor, a Dutch professor, a slang-slinging Texan, and the part-human, part-vampire woman with the man's brain, Mina Harker—they enlist both information and technology. To do so, however, they will also have to expand their circle of *fact*. For who "in the midst of our scientific, matter-of-fact nineteenth century" would have believed that vampires could exist? Professor Van Helsing makes it clear that the assertion that "there are such beings as vampires" is only a statement "until such time as that fact thunder[s] on [the] ear." It must become fact as all facts do, through a complex series of alliances between human and nonhuman actors. "The proof of our own unhappy experience" must combine with "the teachings and the records of the past"—the latter in the forms of "traditions and superstitions"—in order to provide "proof enough for sane peoples." These, with the "resources of science" also included on the hunters' side of the tally, make rather strange bedfellows.[73]

But not unheard of. For the late nineteenth century is rather on the fence about whether a variety of traditions and superstitions might not make good allies in the process of scientific fact making. *Dracula* stages these fact-making anxieties, the prevailing uncertainty about which things will turn up fact, which as mere artifact. Though many physicists disparaged activities such as spiritualism and mesmerism as what we might call (borrowing Maxwell's phrase) "old bosh new dressed," others took seriously the idea that these phenomena should be subject to rigorous treatment by the new physics. Explanations would inevitably be forthcoming, since "the Victorian period was in many ways the heyday of alternative scientific practices." Ether theory in particular seemed to promise connections between the physical world and the phenomena of life and death: "A whole order of forces was waiting to be discovered in the ether." Odism, mesmerism, and spiritualism seemed ripe for explanation through operations of electricity and magnetism or as-yet undiscovered forces like these. Though many scientists worried that such investigations could lead to nothing but charlatanry, and many spiritualists understood their practice as diametrically opposed to the excessive materialism of science, there were nonetheless notable crossovers. "Oliver Lodge—almost the high priest of ether physics" actually managed to publish in *Nature* ("somewhat to his surprise") the positive results of his own experiments on telepathy. And differences between Crookes and Varley's photographs of the ghost Katie King and Roentgen's 1895 X-rays of the human hand were not immediately obvious. Roentgen, it seemed, had indeed discovered a new force in the ether, or at least, a new mysterious form of radiation. As did Marie and Pierre Curie with their researches on uranium and their discoveries of polonium and radium. Such discoveries marked at once a painstaking outgrowth of ether physics, a vindication of claims that it would explain the mysteries of the universe, as well as an articulation of its limits and the beginning of its end. Radioactivity itself "seemed to violate some of the most hallowed principles of physics"—radioactive bodies apparently creating energy, a violation of the first law.[74]

In these early days, it was very hard to see who or what among odism, mesmerism, Crookes, Varley, the Curies, radium, Oliver

Lodge, telepathy, Katie King, Roentgen, uranium, polonium, and X-rays would survive and be useful allies in the fact-building business. Throw Maxwell's demon into the mix, and such distinctions become even harder to make. Fantasies of circumventing the second law abound and in ways that continue to narrow the fine line between the scientific and the spiritual. If we return, for instance, to Stewart and Lockyer, we find a wish to avoid a "purely materialistic view of life" wherein we can "predict all future motions of the living being." They are at great pains to hypothesize an alternative that would nonetheless be consistent with the laws of thermodynamics. To this end, they posit "an independent principle" that may direct the living being: "May not the living being be an organization of infinite delicacy, by means of which a principle, in its essence distinct from matter, by impressing upon it an infinitely small amount of directive energy, may bring about perceptible results?" Certain that "our bodies are machines of exquisite delicacy," they hold that "infinite delicacy" is not inconceivable. Under those circumstances an amount of "directive energy"—the unit of free will, a choice—may make a difference. To avoid violating known laws, however, they insist that the amount of directive energy involved must be infinitesimally small, far smaller than our capacity to measure. So in creating space for free will, they imagine an ideal machine. And in imagining an ideal machine, Stewart and Lockyer also suggest that there may be things in the universe beyond our capacity to explain. Among these is "a Supreme Intelligence [that] without interfering with the ordinary laws of matter, pervades the universe, exercising a directive energy capable of comparison with that which is exercised by a living being." While Stewart and Lockyer thus suggest that God may exist outside but nonetheless consistent with the known laws of matter, others seek to extend similar courtesies to the "so-called spiritual manifestations" in which Stewart and Lockyer "need scarcely . . . inform the reader that [they] do not believe."[75]

Stewart and Lockyer, moreover, develop their notion of "infinite delicacy" by imagining machines that present analogies to living beings, notably a gun with a hair trigger, such that a "small expenditure of energy" may explode a magazine (on the other side of the Atlantic), or "even win an empire."[76] *Dracula* too stages the convergence of empire building with fact building. Indeed, if we revisit Bruno Latour with *Dracula* on the brain, even Latour's prose seems to suggest how much these two have in common. "The problem of the fact builder," he says, is "how to spread out in time and space." In both cases, "we need to do two things at once: *to enrol others* so that they participate in the construction of the fact" (or, we add, of the empire), and "*to control their behaviour* in order to make their actions predictable."[77] Fact builders worry that users may transform statements in unhoped-for directions. *Dracula* worries that the empire might strike back,[78] that figures like Dracula or Mina might not behave properly as sources of energy, conduits of information, or well-behaved interpreters of fact. And so both empire builders and fact builders attempt to reformulate the interests of the groups they wish to enlist—what Latour calls *translation*: "Translating interests means at once offering new interpretations of these interests and channeling people in new directions."[79]

Dracula makes it abundantly clear that desire matters. New vampires *want* to do vampiric things. How is a human empire to compete? And so *Dracula* stages the act of translation. For Mina uncodes not only in the act of typewriting, but also in interpreting the meanings of her own knowledge and abilities.

From the very first, when she reveals that she is practicing shorthand and typing, to the moment she asks to be killed rather than become a vampire and be "leagued with your enemy against you," Mina translates her interests as the interests of the group and the text that ultimately enroll her. Her secretarial work is important not because it might enable, say, financial independence, but because it will make her "useful to Jonathan."[80] She even interprets her knowledge of Dracula's homeland as evidence of her allegiance to her husband's interests, telling Van Helsing that of course she knows the way; hasn't "my Jonathan [travelled it] and [written] of his travel?" Apparently she knows the way because she loves her husband, not because she feels the lure of the vampire. The meanings of her energies as of her facility with information reside in the alliances these signal. Still, anxiety lingers that her allegiances may lie elsewhere, that "some new guiding power" may be operating on her. And during the significant pause before she translates, Mina's knowledge of Dracula's homeland resonates as suspiciously vampiric.[81] It is a pause that raises clear questions about Mina's access to information: What does that access mean? What interests—whose interests—does it serve?

Follow the blood. In *Dracula* "as in the regime of information, struggles for possession yield to issues of access"[82]—to issues, I would add, of circulation of energy, information, and blood. "Indeed, the text orchestrates a symmetry of blood and information flow."[83] Blood is Latour's "black box"—not the stable object whose inner workings we needn't even question, but that transformable object that changes as it gets passed from hand to hand, whose meanings, like those of any statement, change with subsequent users. It reminds us, as Mina's participation reminds the vampire hunters, that "there is no such thing as passive transmission—invariably intelligent knowledge is involved."[84] As Dracula, the vampire hunters, and all the text's readers engage in a struggle to control its flow and its meanings, blood attaches to race and nation, reproduction and nutrition, energy and information, just to name a few. And as the meanings of blood thus prove fluid, new meanings call forth the old. Scientific facts reveal their closeness to artifacts. New meanings of *chaos* evoke early ones, prior to its close association with entropy, such as the late sixteenth-century use of *chaos* for the "primordial mass out of which God created the universe."[85] Similarly likening the flow of blood to that of energy evokes a time when caloric functioned so compellingly as a figure for heat that it was thought to be real. In Maxwell's equations, fluids function as a compelling metaphor for electricity and magnetism to this day. If in *Dracula* blood serves to remind us of the fluid properties of energy and information—those measurable, usable, transferable, transformable, and substantive qualities—it also suggests the lingering power of thermopoetic metaphor. Neither the vampire nor the demon, neither ether nor caloric nor a host of other such imponderables, need be real to be compelling.

Epilogue

Before I came here I was confused about this subject. Having listened
to your lecture I am still confused. But on a higher level.

—Enrico Fermi

Ignorance of the laws of thermodynamics is no excuse.

—Andrew M. Rappe (on the necessity of eating cold latkes)

Throughout *ThermoPoetics*, I have considered physics and fiction
as part of a larger conversation, working out similar concerns,
using overlapping methods, and collaborating in the shaping
and circulation of scientific fact. I have focused on a particularly
solid scientific object—energy—whose status as object we rarely
doubt and whose origins we thus rarely consider. We know
energy, what it can do, what it can't. The facts that go along
with it are so firmly established that we call them laws. Undoubtedly, I have chosen this set of issues because energy matters to
me, to us, now. "Energy," says Nobel Prize–winning physicist
Walter Kohn, "is one of the make-or-break challenges of our
times."[1] As we worry about where we are going in our thinking
about energy, it seems natural to consider where we're coming
from. What energy means builds on what it has meant. And

vice versa; our concerns have an anachronistic effect. They change the shape of Victorian literature, making it look at least as concerned with energy as it has been with evolution or sexuality or colonialism or any of the other issues that have preoccupied its readers. Because all of these concerns are intertwined as part of the thermopoetic fabric we continue to weave, energy turns out to be entangled with each, taking its shape from and shaping its cultural environment.

Yet this doesn't make it one bit less factual. I have never found the constructed nature of facts to be an indictment of their factness—any more than I would think a table less of a table because I knew that it was made through processes I could understand. Nor do I "doubt the fixity of tables" because I know them to be mostly empty space.[2] As far as I'm concerned, understanding is very different from undoing. Analysis is not destruction.

So if energy is still an object and the laws of thermodynamics are still facts, even though we have a better idea of how they were made—of the human baggage that went into their early and contentious formation—then why does it matter? Does it matter how we got here as long as we are here? I like to think so. If our concerns change the face of the Victorians, they may well return the favor. Knowing something about the Victorian conversation that did so much to produce the facts of energy can show us ourselves from a different perspective. It can help to raise questions about what we mean by useful work, how we distinguish order from disorder, and why we want to. And it can help us see that how we answer these questions depends very much on what we mean by *we*. Are we scientists? Humanists? A nation, a species, a planet? And what would it mean to open the system? As we find ourselves in a strikingly analogous situation to the Victorians, worrying that our own favorite sources of energy are running out distressingly quickly, it becomes clear that we are still part of the same conversation, wrestling with the same worrying contradictions, trying to decide who or what to believe and how to maintain some kind of educated optimism, as we struggle to build a better engine. For as energy makes its inexorable way toward heat, we cannot stop it. All we can hope to do is get in the way. And we know this will require not only the collective effort of scientists and humanists and readers and thinkers of all types, but also the cooperation of nonhuman actors, like the sun and the wind, photovoltaics and ferroelectrics.[3]

Appendix

Report on Tait's Lecture on Force:—B.A., 1876[1]
James Clerk Maxwell

Ye British Asses, who expect to hear
      Ever some new thing,
I've nothing new to tell, but what, I fear,
      May be a true thing.
For Tait comes with his plummet and his line,
      Quick to detect your
Old bosh new dressed in what you call a fine
      Popular lecture.

Whence comes that most peculiar smattering,
      Heard in our section?
Pure nonsense, to a scientific swing
      Drilled to perfection?
That small word "Force," they make a barber's block,
      Ready to put on
Meanings most strange and various, fit to shock
      Pupils of Newton.

Ancient and foreign ignorance they throw
      Into the bargain;
The shade of Leibnitz mutters from below
      Horrible jargon.
The phrases of last century in this
      Linger to play tricks—
Vis Viva and *Vis Mortua* and *Vis*
      Acceleratrix:—

Those long-nebbed words that to our text books still
      Cling by their titles,

And from them creep, as entozoa will,
　　　　Into our vitals.
But see! Tait writes in lucid symbols clear
　　　　One small equation;
And Force becomes of Energy a mere
　　　　Space-variation.

Force, then, is Force, but mark you! not a thing,
　　　　Only a Vector;
Thy barbèd arrows now have lost their sting,
　　　　Impotent spectre!
Thy reign, O Force! is over. Now no more
　　　　Heed we thine action;
Repulsion leaves us where we were before,
　　　　So does attraction.

Both Action and Reaction now are gone.
　　　　Just ere they vanished,
Stress joined their hands in peace, and made them one;
　　　　Then they were banished.
The Universe is free from pole to pole,
　　　　Free from all forces.
Rejoice! ye stars—like blessed gods ye roll
　　　　On in your courses.

No more the arrows of the Wrangler race,
　　　　Piercing shall wound you.
Forces no more, those symbols of disgrace,
　　　　Dare to surround you:
But those whose statements baffle all attacks,
　　　　Safe by evasion,—
Whose definitions, like a nose of wax,
　　　　Suit each occasion,—

Whose unreflected rainbow far surpassed
　　　　All our inventions,
Whose very energy appears at last
　　　　Scant of dimensions:—
Are these the gods in whom ye put your trust,
　　　　Lordlings and ladies?
The hidden potency of cosmic dust
　　　　Drives them to Hades.

While you, brave Tait! who know so well the way
　　　　Forces to scatter,
Calmly await the slow but sure decay,
　　　　Even of Matter.

Notes

Prologue

1. Iwan Morus notes William Whewall's coining this term, as well as *scientist*, both in the 1830s (Morus, *When Physics Became King*, 6, 53).

2. Stewart and Lockyer, "Sun as a Type," 319.

3. Smith, *The Science of Energy*, 150.

4. For an excellent discussion of the history of this science, in all of its fascinating social complexity, see Smith, *The Science of Energy*.

5. Clausius, *The Mechanical Theory of Heat with Its Application to the Steam-Engine and to the Physical Properties of Bodies*, 365.

6. Asimov and Shulman, *Isaac Asimov's Book of Science and Nature Quotations*, 75.

7. Chapple, *Science and Literature in the Nineteenth Century*, 44.

8. *Husbands and Wives* (1992). Cited in Wikiquote at http://en .wikiquote.org/wiki/Transwiki:Quotes_&_humor_(thermodynamics).

9. Spencer, *First Principles*, 165.

10. "Nowhere in Dickens's immense literary corpus is the concern with science more profound than in *Household Words* and *All the Year Round*. Dickens wanted science accessible to the general reader and citizen" (Nixon, "'Lost in the vast worlds of wonder,'" 291).

11. "The Chemistry of a Candle," 440–441.

12. Clausius, *Mechanical Theory of Heat*, 357.

13. From a letter to George and Thomas Keats in December of 1817. (*Letters of John Keats*, 43).

14. Stoker, *Dracula*, 227; Tennyson, *In Memoriam*, stanza 40.

15. Stewart and Lockyer, "Sun as a Type," 322.

16. Schneider and Sagan, *Into the Cool*, 6–8.

17. Hayles, *Chaos Bound*, 22.

Introduction

1. This observation finds its way into English in *Hermes*, an influential volume of contemporary literature and science studies, in an essay titled "Turner Translates Carnot," cherished by those with a particular interest in energy physics (Serres, *Hermes: Literature, Science, and Philosophy*, 57).

2. Serres, *Hermes*, 56–57.

3. Dickens himself, "not averse to using science for personal references, once describ[ed] himself at work on *David Copperfield*," not as an engineer, but "as a steam engine" (Nixon, "'Lost in the vast worlds of wonder,'" 273).

4. Latour, *Science in Action*, 132–144.

5. Darwin, *Origin of Species*, 1st ed., chap. 14.

6. Cited in Campbell and Garnett, *The Life of James Clerk Maxwell*, xv.

7. Joule, *The Scientific Papers*, 266–267.

8. Zencey, "Entropy as Root Metaphor," 185; Beer, *Open Fields*.

9. Le Guin, *The Left Hand of Darkness*, xvi.

10. Though Darwin makes several such additions between the first and sixth editions of his *Origin of Species*, the most striking occur in chapter 4, "Natural Selection; *Or the Survival of the Fittest*." I am thinking in

particular of the claim that "Man can act only on external and visible characters: Nature, *if I may be allowed to personify the natural preservation or survival of the fittest*, cares nothing for appearances, except in so far as they are useful to any being." Shortly afterward, he observes that "it may *metaphorically* be said that natural selection is daily and hourly scrutinising, throughout the world" (Darwin, *Origin of Species*, 6th ed., chap. 4). My italics indicate text added in later editions, even to the chapter title.

11. Gillian Beer further identifies Darwin's debt to the example of Charles Dickens, in "the organization of *The Origin of Species* . . . with its apparently unruly superfluity of material gradually and retrospectively revealing itself as order" as in a Dickens novel (Beer, *Darwin's Plots*, 6).

12. Hayles, *Chaos and Order*, 19.

13. This quote is attributed to Enrico Fermi, who was the 1938 Nobel Prize winner in physics; cited by Wikiquote at http://en.wikiquote.org/wiki/Enrico_Fermi.

14. Snow, *The Two Cultures and the Scientific Revolution*, 16.

15. Such numbers have led to Al Gore's "challenge to our nation to commit to producing 100 percent of our electricity from renewable energy and truly clean carbon-free sources within 10 years." (The prepared text of Al Gore's speech on renewable energy, delivered on July 17, 2008, at Constitution Hall in Washington, D. C., can be found at npr.org.)

16. Thomson, "On the Age of the Sun's Heat," 393.

17. Hayles, *Chaos and Order*, 19

18. Blake's term comes from an 1802 letter to Thomas Butts, cited in Whitworth, *Einstein's Wake*, 1.

19. English literature is a relatively new discipline: "In point of fact, University College, London, or as it was called until 1836, 'the University of London' . . . was the first English university institution to introduce the study of English literature, appointing a professor of

English language and literature when it opened in 1828. The emphasis, however, was very heavily on the study of language." (Bacon, "English Literature Becomes a University Subject," 592).

20. Everitt, "Maxwell's Scientific Creativity," 92.

21. Le Guin, *The Left Hand of Darkness*, xvi.

22. Since Thomas Kuhn's influential "Energy Conservation as an Example of Simultaneous Discovery," historians of science such as Stephen Brush, P. M. Harman, and Anson Rabinbach have paid increasing attention to the relation between Victorian thermodynamics and culture more broadly construed. M. Norton Wise and Crosbie Smith's biography of Lord Kelvin, *Energy and Empire*, and Smith's cultural history, *The Science of Energy*, have been extremely important in bringing the latest in scholarly methodology to the historical treatment of Victorian energy physics. These studies represent the kind of work legitimizing Iwan Rhys Morus's efforts to do what once seemed impossible: to produce in *When Physics Became King* an "unashamedly cultural history," one that seeks to look at the big picture of the cultural development of physics (4).

23. "Reading itself proves to have a history," James Secord observes in his introduction to *Victorian Sensation*, and he proceeds to "plac[e] reading at the center of a history" (Secord, *Victorian Sensation*, 3, 4).

24. See, for example, James Paradis and Thomas Postlewait's *Victorian Science and Victorian Values* and Geoffrey Cantor and Sally Shuttleworth's *Science Serialized: Representations of the Sciences in Nineteenth-Century Periodicals*.

25. See especially Peter Allan Dale's *In Pursuit of a Scientific Culture*, in which he devotes a chapter or so to George Eliot's interest in the "higher physics," and Michael Whitworth's lovely discussion of heat, entropy, and especially "dissipation" in his chapter on *The Secret Agent* in *Einstein's Wake*.

26. Levine, *Darwin and the Novelists*, 156; Beer, *Open Fields*, 219.

27. Darwin, *The Descent of Man*, chap. 6.

Chapter 1

1. Stoppard, *Arcadia*, 65.

2. Cited in Stoppard, *Arcadia*, 79.

3. Latour, *Science in Action*, 132–144.

4. Not having reached that point in reading literature and science, wherein a "literary 'theory' falls into disuse not because it has been refuted but because its assumptions have become so visible to its practitioners that it can no longer effectively create the illusion that it is revealing something about the text that is intrinsically present, independent of its assumptions" (Hayles, *Chaos Bound*, 36), I will proceed to treat *In Memoriam* as if it would generate the laws of thermodynamics independent of the investment of the reader. I will leave it to subsequent generations of scholars to demonstrate how bound up in current disciplinary and cultural assumptions such a reading is. Undoubtedly, *In Memoriam* yields such a reading only under specific pressures to do so, but yield such, it nonetheless does.

5. Adams, "Woman Red in Tooth and Claw: Nature and the Feminine in Tennyson and Darwin," 7.

6. Hallam Tennyson, cited in Mattes, *In Memoriam*, 88n26.

7. Cited in Secord, *Victorian Sensation*, 9.

8. Kuhn, *The Structure of Scientific Revolutions*.

9. Susan Gliserman has been extremely important in revitalizing interest in the "cultural exchange" between Victorian science and *In Memoriam*, in "Early Science Writers and Tennyson's 'In Memoriam': A Study in Cultural Exchange." Tess Coslett revisits this aspect of Tennyson's thought, remarking "the way his approach corresponds to that of the scientific writers" and finding in "the vast length of the poem, as opposed to earlier elegies" the gradualism that shapes Lyell's geology (*The Scientific Movement and Victorian Literature*, 43, 47). Isobel Armstrong, who observes similarly that "*In Memoriam* . . . uses the myth of geology structurally as well as absorbing its language," complicates

the geological narrative as she reads "the double poem" created out of the tension between two opposing geological worldviews ("Tennyson in the 1850s: From Geology to Pathology—*In Memoriam* (1850) to *Maude* (1855)," 102, 110). Howard W. Fulweiler places Tennyson's own distinctive theory of evolution in serious scientific conversation with Darwin's and Lyell's ("Tennyson's *In Memoriam* and the Scientific Imagination"). James Eli Adams adds gender to the mix, exploring how "Tennyson's personification of nature . . . suggests that evolutionary speculation also rendered newly problematic a deeply traditional and comforting archetype of womanhood" ("Woman Red in Tooth and Claw," 7). And Jacob Korg, moving away from geology and biology, argues that Tennyson "was more receptive to astronomy than any other poet of his time" ("Astronomical Imagery in Victorian Poetry," 143).

10. Cited in Gibson, "Behind the Veil," 61.

11. Kuhn, *The Structure of Scientific Revolutions*, 66.

12. Peterfreund, "The Re-Emergence of Energy in the Discourse of Literature and Science," 24.

13. Thomson, "On the Dissipation of Energy," 316.

14. Rankine, *Miscellaneous Scientific Papers*, 209.

15. Harman, *Energy, Force, and Matter*, 45; Clarke, "Allegories of Victorian Thermodynamics," 73.

16. Tennyson, *In Memoriam*, stanzas 40, 113.

17. Joule, *The Scientific Papers*, 271.

18. Though it is beyond the scope of this book to trace the influence of Kantian thought on either Tennyson or physics, one aspect of their mutual concerns is particularly relevant here: the emphasis on a unified theory of nature. One of the ways this Kantian idea passes into English science is through Coleridge to his friend, the prominent chemist Humphry Davy (Brush, *The Temperature of History*, 20). Similarly, Laplace's *Treatise on Celestial Mechanics* not only elaborates the nebular hypothesis so resonant in *In Memoriam*, but also insists that heat, light,

and electricity as well as mechanical phenomena can and should be explained by a single, unifying mechanism (Harman, *Energy, Force, and Matter*, 15–19).

19. I am, in my own fashion, continuing to elaborate Latour's "diffusion model." Latour, of course, does a much more thorough job in *Science in Action* (see pp. 132–144).

20. Latour, *Science in Action*, 133.

21. Levine, *Darwin and the Novelists*, 157.

22. Thomson, "On the Age of the Sun's Heat," 391–392.

23. Clausius, *Mechanical Theory of Heat*, 365.

24. Cited in Smith, *The Science of Energy*, 168.

25. Many thanks to my colleague Russel P. Kauffman, who has helped immeasurably in the articulation of the distinction between physical and popular usage.

26. Tennyson, *In Memoriam*, stanza 125.

27. Tennyson, *In Memoriam*, stanza 56.

28. Beer, *Open Fields*, 228.

29. Wells, *"The Time Machine" and "The War of the Worlds,"* 98.

30. Thomson, "On the Age of the Sun's Heat," 393.

31. Tennyson, *In Memoriam*, stanza 3.

32. Cited in Campbell and Garnett, *The Life of James Clerk Maxwell*, xv.

33. Tennyson, *In Memoriam*, stanza 69.

34. Tennyson, *In Memoriam*, stanza 55.

35. Thus in works such as Lewes's essay "Animal Heat" through Rosa's "The Human Body as Engine," we see repeated attempts to explain not only the body but also human psychological, sexual, and social behavior

through the principles of thermodynamics. Anson Rabinbach's chapter "The Political Economy of Labor Power" in *The Human Motor: Energy, Fatigue, and the Origins of Modernity* directly addresses the social implications of energy conservation, and his theme of fatigue suggests the second law throughout. Cynthia Russett's chapter "The Machinery of the Body" in *Sexual Science: The Victorian Construction of Womanhood* considers the deployment of thermodynamic principles, especially in the medical regulation of gender.

36. Rankine, *Miscellaneous Scientific Papers*, 209, 212.

37. Dale, *In Pursuit of a Scientific Culture*, 130, 132.

38. Tennyson, *In Memoriam*, stanzas 22, 34.

39. Stewart and Lockyer, "Sun as a Type," 323.

40. Tennyson, *In Memoriam*, stanzas 50, 61, 125.

41. Meyers, "Nineteenth-Century Popularizations of Thermodynamics and the Rhetoric of Social Prophecy," 308.

42. Dale, *In Pursuit of a Scientific Culture*, 6.

43. Morus, *When Physics Became King*, 58.

44. Brush, *The Temperature of History*, 20, 29.

45. Clarke, *Energy Forms*, 21.

46. Viswas, *Tennyson's Romantic Heritage*, 4.

47. Coleridge, "Religious Musings," lines 110–111.

48. Peterfreund, "Re-Emergence of Energy," 24, 34, 40–41.

49. Tennyson, *In Memoriam*, stanza 40.

50. Beer, "Tennyson, Coleridge, and the Cambridge Apostles," 1–2.

51. Peterfreund, "Re-Emergence of Energy," 40–41.

52. Tennyson, *In Memoriam*, stanza 122.

53. Dale, *In Pursuit of a Scientific Culture*, 7.

54. Peterfreund, "Re-Emergence of Energy," 43.

55. Dale, *In Pursuit of a Scientific Culture*, 135.

56. Richard Butts illustrates this shift in the nineteenth century from *univocal* to *equivocal* modes of thought—the former insistent that scientific theories can and should reveal real truths about nature; the latter understanding scientific hypotheses as symbolic constructs. For Butts, the analogical mode, characteristic of both Tennyson and Tyndall, represents a compromise between these extremes. From a slightly different angle, Patricia O'Neill calls attention to "Tyndall's sympathy with the Romantic poet's task to humanize the experience of nature." See Butts, "Languages of Description and Analogy in Victorian Science and Poetry"; O'Neill, "Victorian Lucretius: Tennyson and the Problem of Scientific Romanticism," 106.

57. Joule, *The Scientific Papers*, 266–267.

58. See Zencey, "Entropy as Root Metaphor."

59. Clarke, "Allegories of Victorian Thermodynamics," 74.

60. Tennyson, *In Memoriam*, stanzas 118, 84, 84.

61. Carnot, "Reflections on the Motive Power of Fire," 10.

62. Tennyson, *In Memoriam*, stanza 118.

63. Tennyson, *In Memoriam*, stanzas 50, 62, 47 (emphasis mine).

64. Tennyson, *In Memoriam*, Epilogue.

65. Butts, "Languages of Description and Analogy in Victorian Science and Poetry," 206–207.

66. Tennyson, *In Memoriam*, stanza 82.

67. Tennyson, *In Memoriam*, stanza 41.

68. Tennyson, *In Memoriam*, stanza 85.

69. Tennyson, *In Memoriam*, stanza 130.

70. Wordsworth, "Ode on Intimations of Immortality from Recollections of Early Childhood," stanza 10.

71. Tennyson, *In Memoriam*, stanza 130.

72. Maxwell, *The Scientific Papers of James Clerk Maxwell*, 322.

73. Tennyson, *In Memoriam*, Epilogue.

74. Tennyson, *In Memoriam*, stanza 56.

75. Joule, *The Scientific Papers*, 266–267.

76. Thomson, *Mathematical and Physical Papers*, 115–117.

77. Latour, *Science in Action*, 133.

78. Thomson, *Mathematical and Physical Papers*, 102n.

79. Thomson, *Mathematical and Physical Papers*, 116–117.

80. Thomson, *Mathematical and Physical Papers*, 118–119n (emphasis mine).

81. Joule, *The Scientific Papers*, 188–189.

82. Joule, *The Scientific Papers*, 269, 272.

83. Tennyson, *In Memoriam*, 75.

84. Tennyson, *In Memoriam*, 273.

85. Thomson, *Mathematical and Physical Papers*, 511. Such is the power of the notion of conservation that even God is often said to have created the universe out of something. As Alice Jenkins observes, in a tradition that does not predominate in the early nineteenth century, but that touches *Frankenstein* and echoes "the most favourable descriptions . . . in *Paradise Lost*," chaos has been repeatedly treated not as a site of destruction or disintegration, but rather as a "storehouse of matter awaiting organization" (Jenkins, *Space and the "March of Mind*," 216).

86. Craft, "'Descend, Touch, and Enter': Tennyson's Strange Manner of Address," 85, 88. Deploying his own thermodynamic metaphor, Craft also notes that "the elegy negotiates its problematic desire less by a centering of its warmth than by the dispersion of its bliss" (p. 85).

87. Nunokawa, *"In Memoriam* and the Extinction of the Homosexual."

88. Tennyson, *In Memoriam*, stanzas 56, 69, 55, 111.

89. Gliserman, "Early Science Writers," 279.

90. Tennyson, *In Memoriam*, stanza 35.

91. Tennyson, *In Memoriam*, stanza 123.

92. Tennyson, *In Memoriam*, stanza 55.

93. Tennyson, *In Memoriam*, stanza 118.

94. Fulweiler, "Tennyson's *In Memoriam* and the Scientific Imagination," 315, 313.

95. Tennyson, *In Memoriam*, Epilogue (emphasis mine).

Chapter 2

1. From Tyndall's "Belfast Address."

2. In response to Mary Somerville's *On the Connexion of the Physical Sciences*, Maxwell himself observed this "widespread desire to be able to form some notion of physical science as a whole" (cited in Chapple, *Science and Literature in the Nineteenth Century*, 46).

3. Cited in Smith, *The Science of Energy*, 185. Thomson and Tait represent what Crosbie Smith calls the "North British" school. Smith's chapter "North Britain *versus* Metropolis" discusses in depth these competing schools of the developing science of energy, especially as regards their relations to religion and scientific naturalism, as well as contention over scientific authority.

4. Dale, *In Pursuit of a Scientific Culture*, 14.

5. From *Principles of Biology* (1864).

6. Spencer, *First Principles*, 484, 483.

7. Spencer, *First Principles*, 491, 484, 495.

8. In his discussion of energy concepts in the context of Western allegory, Bruce Clarke observes that "allegory and science are both embedded in overlapping textual and historical traditions that discipline their operation as technologies of meaning. . . . Allegory has always offered science extended metaphors, flexible heuristic frameworks for investigation and reformulation" (*Energy Forms*, 22).

9. Smith, *The Science of Energy*, 42, 219.

10. Unless otherwise specified, page references are to the 1979 Woodbridge Press edition.

11. Sponsored by the English Department at San Jose State University, www.bulwer-lytton.com.

12. Sinnema, introduction to Bulwer-Lytton, *The Coming Race*, 13, 9–10.

13. Bulwer-Lytton read a chapter of Spencer's first book, *Social Statics* (1851), and commented favorably, though he ultimately disagreed with Spencer's thinking on "national education" (cited in Duncan, *The Life and Letters of Herbert Spencer*, 78).

14. Bulwer-Lytton, *The Coming Race*, 27.

15. Suggestive titles along these lines include Anne-Julia Zwierlein, "The Evolution of Frogs and Philosophers: William Paley's *Natural Theology*, G. H. Lewes's *Studies in Animal Life* and Edward Bulwer-Lytton's *The Coming Race*," and Nancy L. Paxton, *George Eliot and Herbert Spencer: Feminism, Evolutionism, and the Reconstruction of Gender*. Peter Sinnema observes that the novel "discerns an enigma at the heart of evolutionary theory" (introduction to Bulwer-Lytton, *The Coming Race*, Broadview Press ed., 24).

16. Bulwer-Lytton also shares with Spencer the tendency to extend Darwinian theory to social arguments: "The argument of *The Coming Race* is that this social evolution, instead of leading to a higher civilization, is in fact a degenerative process" (Williams, *Notes on the Underground*, 128). For us, the degenerative nature of this process speaks to concerns regarding the second law.

17. Spencer, *First Principles*, 494.

18. Spencer, *First Principles*, 358–359.

19. Bulwer-Lytton, *The Coming Race*, Broadview Press ed., 169.

20. Bulwer-Lytton, *The Coming Race*, 52, 57.

21. Bulwer-Lytton, *The Coming Race*, 8.

22. Bulwer-Lytton, *The Coming Race*, 20–21.

23. Clarke identifies Vril predominantly with another physicist's efforts at unification: "It appears that Lytton modeled vril, at least in a round-about way, on the same all-pervasive luminiferous or electromagnetic ether through which Maxwell had recently produced a unified field theory of radiant energies" (*Energy Forms*, 49).

24. Bulwer-Lytton, *The Coming Race*, 29.

25. Spencer, *First Principles*, 196–199.

26. Bulwer-Lytton, *The Coming Race*, 25–26.

27. Spencer, *First Principles*, 493.

28. Spencer, *First Principles*, 448.

29. Spencer, *First Principles*, 457–459.

30. Bulwer-Lytton, *The Coming Race*, 26.

31. Bulwer-Lytton, *The Coming Race*, 67, 27.

32. Bulwer-Lytton, *The Coming Race*, 67.

33. Bulwer-Lytton, *The Coming Race*, 28.

34. Bulwer-Lytton, *The Coming Race*, 67.

35. Bulwer-Lytton, *The Coming Race*, 28.

36. Spencer, *First Principles*, 346.

37. Spencer, *First Principles*, 459.

38. Bulwer-Lytton, *The Coming Race*, 19, 4, 6.

39. Bulwer-Lytton, *The Coming Race*, 64, 20.

40. Bulwer-Lytton, *The Coming Race*, 58.

41. There is a nice discussion of how metaphor is made literal in James and Mendelsohn, *The Cambridge Companion to Science Fiction*, 5.

42. Bulwer-Lytton, *The Coming Race*, 58.

43. Joule, *The Scientific Papers*, 266–267.

44. Bulwer-Lytton, *The Coming Race*, 58.

45. These aspects of *The Coming Race* suggest how it may be read as an occult novel. Energy concepts may have appealed to Bulwer-Lytton because they enable this move from the materialist focus of much modern science, to more spiritual concerns. As Rosalind Williams observes, regarding this and other subterranean fictions: "The lower world, is, paradoxically, identified with higher truth" (*Notes on the Underground*, 45–46).

46. Stewart and Tait, *The Unseen Universe or Physical Speculations on a Future State*, 5, 27, 271.

47. Stewart and Tait, *The Unseen Universe or Physical Speculations on a Future State*, 96.

48. Spencer, *First Principles*, 272.

49. Clarke, *Energy Forms*, 66.

50. Spencer, *First Principles*, 497–498.

51. Stewart and Tait's adherence to and interpretation of the Principle of Continuity disallows such inconceivability. To them, "The Law of Continuity . . . simply means that the whole universe is of a piece; that it is something which an intelligent being is capable of understanding" (*The Unseen Universe or Physical Speculations on a Future State*, 270–271).

52. Bulwer-Lytton, *The Coming Race*, 58.

53. Bulwer-Lytton, 26, 31, 25–26.

54. Clarke asserts that "Lytton's fable . . . is first and foremost a plea for imperial fortitude and a cautionary statement about the domestic and overseas perils of 'equilibration,' that is, liberal democraticization." I would suggest that this reading is correct but not complete. Undoubtedly a "conservative or Tory satire of political liberalism in defense of colonial motives" (Clarke, *Energy Forms*, 51, 32), *The Coming Race* must nonetheless be read as profoundly ambivalent about these motives as well as about the repercussions of the British imperial project. At once a projection of British imperialism and of the colonial Other, the Vril-ya also evoke the complications that attach to self-perceived *advancedness*, the "anxiety of reverse colonization," and the guilt that goes with these (Arata, "The Occidental Tourist: *Dracula* and the Anxiety of Reverse Colonization").

55. Spencer, *First Principles*, 162, 165.

56. Spencer, *First Principles*, 459.

57. Dale, *In Pursuit of a Scientific Culture*, 132.

58. Smith, *The Science of Energy*, 180, 183.

59. For a discussion of this exchange and these quotes at greater length, see Smith, *The Science of Energy*, 224.

60. Bulwer-Lytton, *The Coming Race*, 27, 21.

61. Bulwer-Lytton, *The Coming Race*, 61.

62. Bulwer-Lytton, *The Coming Race*, 38–39.

63. Spencer, *First Principles*, 459–460.

64. See "The Reinvention of Newton" in Smith, *The Science of Energy*, as well as the next chapter of this book.

65. Bulwer-Lytton, *The Coming Race*, 19. The Vril-ya themselves may also be taken to represent the threat of the increasingly powerful American empire, contrasted with a perception of Britain's decline. Lillian Nayder observes the novel's anxieties about "America's rise to imperial power" as well as its attempts to "defuse this threat" through

its satirical treatment of the American narrator (p. 215). She further observes that "the Vril-ya embody Bulwer-Lytton's sense that the relation between the colonized and the colonizer has become thoroughly unstable" (p. 217).

66. Bulwer-Lytton, *The Coming Race*, 76.

67. Smith, *The Science of Energy*, 124.

68. Thomson, "On the Age of the Sun's Heat," 393.

69. Bulwer-Lytton, *The Coming Race*, 54.

70. Bulwer-Lytton, *The Coming Race*, 72.

71. Wise, "Time Discovered and Time Gendered in Victorian Science and Culture," 53.

72. Bulwer-Lytton, *The Coming Race*, 72.

73. Bulwer-Lytton, *The Coming Race*, 54.

74. Bulwer-Lytton, *The Coming Race*, 54.

75. Bulwer-Lytton, *The Coming Race*, 49.

76. Bulwer-Lytton, *The Coming Race*, 118.

77. I happened to glance this un-self-consciously thermodynamic assertion on the back of a stranger's T-shirt.

78. Spencer, *First Principles*, 449.

79. Bulwer-Lytton, *The Coming Race*, 64.

80. In Lytton's dedication to Max Müller, Bruce Clarke finds an indication that "Lytton's allegory of energy is also an inversion of solar mythology" (*Energy Forms*, 49).

81. Bulwer-Lytton, *The Coming Race*, 18.

82. In this way, the sun-as-centralized-power-source stands out as particularly emblematic of the structure Edward Said distinguishes as *impe-*

rialist, for its focus on a central ruling power: " 'Imperialism' means the practice, the theory, and the attitudes of a dominating metropolitan center ruling a distant territory; 'colonialism,' which is almost always a consequence of imperialism, is the implanting of settlements on distant territories" (Said, *Culture and Imperialism*, 9).

83. Bulwer-Lytton, *The Coming Race*, 110.

84. Cited in Whitworth, *Einstein's Wake*, 62.

85. Bulwer-Lytton, *The Coming Race*, 115.

86. Spencer, *First Principles*, 461.

87. Cited in Clarke, *Energy Forms*, 67.

88. Spencer, *First Principles*, 337.

89. Spencer, *First Principles*, 415, 328.

90. Bulwer-Lytton, *The Coming Race*, 28.

91. Spencer, *First Principles*, 449, 441, 493, 440.

92. Bulwer-Lytton, *The Coming Race*, 19, 26–27 (emphasis mine).

93. Bulwer-Lytton, *The Coming Race*, 52.

94. Clarke, *Energy Forms*, 52.

95. Bulwer-Lytton, *The Coming Race*, 128.

Chapter 3

1. Bruce Clarke discusses this quality at length, tracing even the relation of the word *entropy* to the word *trope* in "Allegories."

2. Latour, *Science in Action*, 23, 31.

3. Bruno Latour explains these processes very nicely in the "Literature" chapter of *Science in Action*, 21–62.

4. Delivered on September 8, 1876, at the Glasgow meeting of the British Association and published in *Nature* (September 21, 1876, 459–

463). Reprinted in Tait, *Lectures on Some Recent Advances in Physical Science, with a Special Lecture on Force.*

5. Tait, *Lectures*, 344–345.

6. In spite of this effort to establish scientific univocality, Tait is by no means averse to punning with scientific language and even mathematics, as his habit of referring to his friends in mathematical notation associated with thermodynamic quantities strongly suggests. Maxwell himself becomes dp/dt from the equation $JCM = dp/dt$ (see Harman, *Energy, Force, and Matter*, 96–97). The use of the dp/dt notation, moreover, suggests the ascendancy of post-Newtonian, French-inspired mathematics, which became dominant by the end of the nineteenth century. Tait's play on Maxwell's initials thus entered into a tradition of Cantabrigian mathematical punning. Indeed, the Analytical Society, founded in 1811, expressed its commitment to the new French analytical calculus—with its dx/dy (read: D. X. D. Y.) notation replacing the old $\dot{x}$ (read: X-Dot)—as supporting "the Principles of pure D'ism in opposition to the Dot-age of the University" (cited in Morus, *When Physics Became King*, 35).

7. Tait, *Lectures*, 345.

8. Maxwell's poems are included at the end of Campbell and Garnett's *Life*. The "Report" can be found on pages 646–648 as well as in the appendix to this book.

9. Anonymous review of Joule's *Scientific Papers* (cited in Smith, *The Science of Energy*, 1).

10. Tait, *Lectures*, 354.

11. Tait makes these observations in the almost-explicit critique of Tyndall that slips into his lecture on force, where he asserts that "heat and kinetic energy in general are no more *'modes of motion'* than potential energy of every kind (including that of unfired gunpowder) is a *'mode of rest'*" (*Lectures*, 367; original italics).

12. See especially Smith's chapter "Gentleman of Energy" in *The Science of Energy*.

13. Cited in Smith, *The Science of Energy*, 227.

14. Harman, *Energy, Force, and Matter*, 85, 95.

15. Cited in Campbell and Garnett, *Life*, 612–617.

16. Cited in Campbell and Garnett, *Life*, 625–628.

17. This symbol represents the mathematical operation of taking a differential in three dimensions—a vector differential.

18. Cited in Campbell and Garnett, *Life*, 634–636.

19. Campbell and Garnett, *Life*, 634. The author's footnote refers to Tait's discussion of Hamilton's "quaternion operator," which is that same "del" operator mentioned above. It is used to define common properties of a field, properties known as "DIV," "GRAD," and "CURL."

20. The equation in which "Force becomes of Energy a mere space variation" is (in one dimension) $F = dE/dx$, which can be rewritten as $Fdx = dE$. This basically means that, as one puts energy into a system (lifts a weight, compresses a spring, whatever), the amount of energy put in is equal to the force applied multiplied by the distance over which it is applied. When the force is changing, this has to be done as an integral. If the force remains constant, we get the simplified version of the statement, energy = force × distance. Said otherwise, this equation defines force as the spatial rate at which energy is being provided to (or, in the negative, drawn from) a system.

21. Tait, *Lectures*, 365.

22. Tait, *Lectures*, 366.

23. For the details of these efforts, see "Reinventing Newton" in Smith, *The Science of Energy*. While Iwan Morus also notes such efforts to establish Newton as "energy's precursor," he also elaborates an earlier resistance to Newton within the development of nineteenth-century physics, especially to Newton's "grand dictum of *hypothesis non fingo*—'I do not feign hypotheses.'" Similar turning from Newton can also be seen in the shift to analytical calculus (Morus, *When Physics Became King*, 79, 60, 35).

24. Cited in Campbell and Garnett, *Life*, 648.

25. Tait, *Lectures*, 357.

26. Tait, *Lectures*, 356.

27. Tait, *Lectures*, 365.

28. See Iltis, "Leibniz and the *Vis Viva* Controversy."

29. Tait, *Lectures*, 346.

30. Smith, *The Science of Energy*, 211, 225.

31. Probably the "sage of Leipzig" whom Campbell and Garnett foot-note as an alternative to "shade of Leibnitz" (*Life*, 647n). Maxwell-Boltzmann statistics describe the statistical distribution of the energy states, velocities, and so on, of particles in thermal equilibrium. These statistics hold only when quantum mechanical effects are negligible, of course, but that amounts to pretty much all macroscopic phenomena.

32. Smith, *The Science of Energy*, 263.

33. Stewart and Tait, *The Unseen Universe*, 1–2.

34. Stewart and Tait are not alone in such efforts. After Tyndall's 1874 Belfast Address, the periodical press moved from casting him in a gener-ally positive light to portraying him as "an aggressive, dishonest, devious, and distinctly un-British materialist" (Lightman, "Scientists as Materialists in the Periodical Press: Tyndall's Belfast Address," 202).

35. Stewart and Tait, *The Unseen Universe*, xi.

36. Stewart and Tait, *The Unseen Universe*, 27.

37. Latour, *Science in Action*, 133.

38. For a discussion of the impact of revolutionary changes made on French physics, see Morus's chapter "A Revolutionary Science" as well as the section "Engineering France" (*When Physics Became King*, 127–133).

39. Mendoza, introduction to Carnot, "Reflections on the Motive Power of Fire," x–xi.

40. Carnot, "Reflections," 4–5.

41. I am thinking of a notion of discovery that incorporates the meaning of *fact* evoking "the root *facere, factum* (to make or to do)" as in *factory*, as opposed to the more usual sense in which "fact is taken to refer to some objectively independent entity which, by reason of its 'out thereness' cannot be modified at will and is not susceptible to change under any circumstances" (Latour and Woolgar, *Laboratory Life*, 175).

42. Carnot, "Reflections," 4–5.

43. Carnot, "Reflections," 5.

44. Siemens, "A New Theory of the Sun: The Conservation of Solar Energy," 510.

45. "The Sun's Surroundings and the Coming Eclipse," 297.

46. "Age of the Sun and Earth," 323.

47. Lockyer, "What the Sun Is Made Of," 75.

48. Thomson, "On the Age of the Sun's Heat," 393.

49. Munro, "New Sources of Electric Power," 919.

50. Stewart and Lockyer, "Sun as a Type," 320, 322.

51. For a very readable account of self-similarity and the butterfly effect, see James Gleick, *Chaos: Making a New Science*. For a discussion of the ways chaos theory grew out of nineteenth-century energy physics, especially as that growth was enabled by heuristic narratives attached to science, see Hayles, *Chaos Bound*. For more on chaos and Victorian literature, see chapter 6 of this book.

52. Stewart and Lockyer, "Sun as a Type," 257.

53. Stewart and Lockyer, "Sun as a Type," 319–320.

54. Stewart and Lockyer, "Sun as a Type," 319–320.

55. Stewart and Lockyer, "Sun as a Type," 319.

56. Stewart and Lockyer, "Sun as a Type," 319.

57. See chapter 1.

58. Joule, *The Scientific Papers*, 273.

59. See Smith, *The Science of Energy*, 72, 55–56.

60. Joule, *The Scientific Papers*, 273.

61. Smith, *The Science of Energy*, 101.

62. For details on Thomson and Joule's social and religious views, see especially the chapters "From Design to Dissolution" and "Mr Joule of Manchester" in Smith, *The Science of Energy*.

63. Stewart and Lockyer, "Sun as a Type," 327.

64. Stewart and Lockyer, "Sun as a Type," 321.

65. Stewart and Lockyer, "Sun as a Type," 321–322.

66. Cited in Whitworth, *Einstein's Wake*, 68.

67. Smith, *The Science of Energy*, 124.

68. Stewart and Lockyer, "Sun as a Type," 322.

69. Stewart and Lockyer, "Sun as a Type," 323.

70. Stewart and Lockyer, "Sun as a Type," 322–323.

71. Stewart and Lockyer, "Sun as a Type," 323.

72. Stewart and Lockyer, "Sun as a Type," 324–325.

73. Stewart and Lockyer, "Sun as a Type," 326–327.

74. Carnot, "Reflections," 4.

Chapter 4

1. To Chevalier de Ouis, 1814, etext.virginia.edu/jefferson/quotations/jeff1770.html.

2. Cited in Clarke, *Energy Forms*, 67.

3. Latour, *Science in Action*, 25.

4. Latour, *Science in Action*, 43.

5. Dickens, *A Tale of Two Cities*, 3.

6. Dickens, *A Tale of Two Cities*, 3–6.

7. Dickens, *A Tale of Two Cities*, 377.

8. George Levine identifies this take on thermodynamics in *Little Dorrit*, whose prose he finds "energetically preoccupied with disorder and loss of energy. . . . It thematizes failure of energy" (*Darwin and the Novelists*, 156).

9. For more on this aspect of Spencer, see chapter 2.

10. Spencer, *First Principles*, 471.

11. Spencer, *First Principles*, 447.

12. Cited in Clarke, *Energy Forms*, 67.

13. Spencer, *First Principles*, 449.

14. Dickens's astuteness regarding human behaviors and institutions has often been contrasted with his relative disinterest in science. But recent scholarship on Dickens has made it clear how much Dickens did engage with the scientific discourses of his day. For an excellent survey of this body of work, see Jude Nixon's " 'Lost in the Vast Worlds of Wonder': Dickens and Science." Gillian Beer's treatment of Dickens in *Darwin's Plots*, and George Levine's in *Darwin and the Novelists*, have made it seem rather natural to discuss Dickens and science—and not just as the dehumanizing force it may seem from, say, *Hard Times*. And scholars like K. J. Fielding and Shu-Fang Lai have made efforts to trace "actualities" indicative of Dickens's engagement with science, among these, the record of his library, his 1860 acquisition of *The Origin of Species*, and his review of Robert Hunt's *The Poetry of Science* (1848), which in turn indicates Dickens's enthusiasm about Robert Chamber's *Vestiges of the Natural History of Creation* (1844).

15. Dickens, *A Tale of Two Cities*, 266, 218, 177–178, 217, 224.

16. Dickens, *A Tale of Two Cities*, 127, 19, 177–178, 236.

17. Dickens, *A Tale of Two Cities*, 108, 229.

18. Dickens, *A Tale of Two Cities*, 217.

19. Levine, *Darwin and the Novelists*, 161. Jude Nixon observes similarly, regarding Dickens's *Dombey and Son*, that "one must pursue the clearly delineated relationship (and not mere analogy) between the life of science and the science of life" (" 'Lost in the Vast Worlds of Wonder,' " 289). Fielding and Lai observe that Dickens "vigorously supported" Robert Hunt's view that "science was imaginative and almost poetic" and himself believed that "we could profit from science because it roused one's imagination as well as showed us the truth" (p. 8).

20. Dickens, *A Tale of Two Cities*, 218.

21. Dickens, *A Tale of Two Cities*, 222.

22. Dickens, *A Tale of Two Cities*, 234.

23. Dickens, *A Tale of Two Cities*, 167.

24. Dickens, *A Tale of Two Cities*, 236.

25. Serres, *Hermes*, 56.

26. Clarke, "Allegories of Victorian Thermodynamics," 74.

27. Dickens, *A Tale of Two Cities*, 229.

28. Dickens, *A Tale of Two Cities*, 108.

29. Dickens, *A Tale of Two Cities*, 107.

30. Dickens, *A Tale of Two Cities*, 239.

31. The first quote comes from *A Tale of Two Cities*, 239. By a meaningless coincidence, the second comes from Smith, *The Science of Energy*, 239.

32. Spencer, *First Principles*, 326–327.

33. Spencer, *First Principles*, 328.

34. Bulwer-Lytton, *The Coming Race*, 71.

35. Dickens, *A Tale of Two Cities*, 343.

36. Spencer, *First Principles*, 337.

37. Dickens, *A Tale of Two Cities*, 106.

38. Dickens, *A Tale of Two Cities*, 114.

39. Dickens, *A Tale of Two Cities*, 240–241.

40. Zencey, "Entropy as Root Metaphor," 188.

41. Dickens, *A Tale of Two Cities*, 285, 124.

42. Dickens, *A Tale of Two Cities*, 244, 246.

43. Dickens, *A Tale of Two Cities*, 124.

44. Cited in Spencer, *First Principles*, 179.

45. Spencer, *First Principles*, 176, 180, 196–197.

46. Spencer, *First Principles*, 465.

47. Dickens, *A Tale of Two Cities*, 376.

48. Dickens, *A Tale of Two Cities*, 372.

49. The English astrophysicist Arthur Eddington coined and popularized this term in the late 1920s.

50. Spencer, *First Principles*, 238.

51. Dickens, *A Tale of Two Cities*, 266, 238.

52. Dickens, *A Tale of Two Cities*, 22–23. Jarvis Lorry's machinelike qualities attach to his profession as an agent of Tellson's bank. Jennifer Fleissner discusses the simultaneous "feminization and mechanization of clerical work" in the late nineteenth century, a move that converges nicely in the semantic confusion wrought by referring to both the woman and her machine as "typewriters" ("Dictation Anxiety," 424–425).

53. Dickens, *A Tale of Two Cities*, 18.

54. Dickens, *A Tale of Two Cities*, 235.

55. Spencer, *First Principles*, 66.

56. Dickens, *A Tale of Two Cities*, 204.

57. Dickens, *A Tale of Two Cities*, 187.

58. Dickens, *A Tale of Two Cities*, 205, 344.

59. Dickens, *A Tale of Two Cities*, 46, 49.

60. Dickens, *A Tale of Two Cities*, 131.

61. Dickens, *A Tale of Two Cities*, 273–274.

62. Dickens, *A Tale of Two Cities*, 263.

63. Dickens, *A Tale of Two Cities*, 273 (emphasis mine).

64. Dickens, *A Tale of Two Cities*, 213, 89.

65. Dickens, *A Tale of Two Cities*, 91.

66. Dickens, *A Tale of Two Cities*, 150, 317.

67. Dickens, *A Tale of Two Cities*, 317.

68. Zencey, "Entropy as Root Metaphor," 193.

69. Dickens, *A Tale of Two Cities*, 74.

70. Dickens, *A Tale of Two Cities*, 85.

71. Dickens, *A Tale of Two Cities*, 150–152.

72. Schneider and Sagan, *Into the Cool*, xi.

73. Dickens, *A Tale of Two Cities*, 301.

74. Dickens, *A Tale of Two Cities*, 352.

75. Dickens, *A Tale of Two Cities*, 375–376.

76. Spencer, *First Principles*, 474.

77. Rutherford recounts his misgivings about announcing his estimates of the earth's age in a 1904 lecture to the Royal Institution at which William Thomson, long since knighted Lord Kelvin, was present. "To

my relief, Kelvin fell asleep, but as I came to the important point, I saw the old bird sit up, open an eye and cock a baleful glance at me! Then a sudden inspiration came, and I said Lord Kelvin had limited the age of the earth, provided no new source was discovered. That prophetic utterance refers to what we are now considering tonight, radium! Behold! The old boy beamed upon me" (cited in Brush, *The Temperature of History*, 43).

78. Spencer, *First Principles*, 474.

79. Levine, *Darwin and the Novelists*, 156.

80. Levine, *Darwin and the Novelists*, 261.

Chapter 5

1. Stewart, *The Conservation of Energy*, vii.

2. From the episode titled "The PTA Disbands," originally aired April 16, 1995. Cited in Maher, "In This House We Obey the Laws of Thermodynamics," 405.

3. See chapter 1.

4. Originally published as a monthly serial from March 1852 to September 1853.

5. Several scholars have noted this novel's engagement with what was to become *entropy*, including Peter Ackroyd, J. A. V. Chapple, and Ann Y. Wilkinson, all of whom are mentioned in Jude V. Nixon's survey.

6. Latour, *Science in Action*, 88.

7. This qualified optimism may be said to characterize Dickens's relationship to science more broadly. According to Adelene Buckland, Dickens understands science as both destructive and productive. In a process that might, for us, recall the kind of transformations to which energy is subject, "Science does not merely 'destroy' fantastical and aesthetic elements of mythology but reconverts them, complete with a sense of wonder and awe, into observationally-verified, scientific sets of facts" (Buckland, " 'The Poetry of Science,' " 681).

8. Working close to my own concerns, Ann Y. Wilkinson, in "From Faraday to Judgment Day," takes up the scientific underpinnings of Krook's spontaneous combustion, locating these in Dickens's connections to the highly esteemed Michael Faraday, leader of the immediately preceding generation of physical scientists. Devoting considerable effort to establishing the validity of reading Dickens in the context of physical science, however, Wilkinson only scratches the surface of *Bleak House*'s thorough and surprisingly optimistic engagement with thermodynamics.

9. Dickens, *Bleak House*, 424.

10. Dickens, *Bleak House*, xxvii.

11. Dickens, *Bleak House*, 434–435. Rosemarie Bodenheimer understands Dickens's defense of spontaneous combustion as "turn[ing] the controversy into a defense of witnessed experience against scientific theory," and the debate between Dickens and Lewes as a clear "contest between two kinds of knowledge: the scientific versus the anecdotal" (*Knowing Dickens*, 10, 9).

12. For the details of their rather lively exchange, see Haight, "Dickens and Lewis on Spontaneous Combustion."

13. Latour illustrates how these moves work in the scientific article in the "Literature" chapter of *Science in Action*. See especially pages 30–38.

14. Cited in Haight, "Dickens and Lewis on Spontaneous Combustion," 55.

15. Dickens, *Bleak House*, 1.

16. Carl Freedman finds "Dickens's London . . . overwhelmingly science-fictional." From the late eighteenth century, London came to look like a city of the future, capable of the kind of "properly science-fictional cognitive estrangement of reality" that allows a reader the imaginative distance to ponder such problems as that of human life in *Frankenstein* or evolution in *The Time Machine* (Freedman, "London as Science Fiction," 259, 252).

17. Tennyson, *In Memoriam*, stanza 69.

18. Dickens, *Bleak House*, 1–2.

19. Dickens, *Bleak House*, 26.

20. The entropic character of Chancery connects to what D. A. Miller identifies as "the ultimate unlocalizability of its operations [which] permits them to be in all places at once" (*The Novel and the Police*, 60). For Miller, the power that Chancery represents complicates notions of discipline based on imprisonment, because it makes "the binarisms of inside/outside, here/elsewhere" on which these notions depend "become meaningless and the ideological effects they ground impossible" (p. 62). Said very simply, if the power of Chancery operates everywhere, it disciplines not just those whom it imprisons, but everyone within reach of its influence.The problematics of inside/outside—which Miller extends to the form of the novel itself—are also central to the problems of engineering, which seek to produce work in spite of the second law, as we shall see.

21. Dickens, *Bleak House*, 2–5.

22. Dickens, *Bleak House*, 110.

23. Dickens, *Bleak House*, 50.

24. Dickens, *Bleak House*, 201.

25. Dickens, *Bleak House*, 328.

26. Dickens, *Bleak House*, 49, 46, 49.

27. Dickens, *Bleak House*, 178–180.

28. Dickens, *Bleak House*, 180.

29. Dickens, *Bleak House*, 141.

30. More on the connection between energy or more especially entropy and information in the next chapter.

31. Dickens, *Bleak House*, 404 (emphasis mine).

32. Dickens, *Bleak House*, 490, 225, 217.

33. Dickens, *Bleak House*, 744.

34. Dickens, *Bleak House*, 772.

35. Dickens, *Bleak House*, 772.

36. I have to acknowledge the completely esoteric allusion to Philip Levine's lovely volume of poetry by the name *What Work Is*.

37. Morus, *When Physics Became King*, 123 (emphasis mine).

38. Dickens, *Bleak House*, 97, 100, 97.

39. Dickens, *Bleak House*, 97, 101.

40. Dickens, *Bleak House*, 99.

41. Dickens, *Bleak House*, 66.

42. Dickens, *Bleak House*, 35, 172, 32.

43. Dickens, *Bleak House*, 41.

44. See prologue.

45. Dickens, *Bleak House*, 35–36.

46. Dickens, *Bleak House*, 622.

47. Esther's warmth may be read in the anger and sexual desire she exhibits, in spite of her outward conformity with conduct-book notions of appropriate female behavior and emotional restraint (Currie, "Against the Feminine Stereotype").

48. Dickens, *Bleak House*, 17, 218.

49. Dickens, *Bleak House*, 39.

50. Robert Tracy observes that "as a housekeeper's narrative, [*Bleak House*] is shaped by a commitment to ordering chaos, and Esther deals with the events of her story as she dealt with Bleak House, by creating a catalogue or inventory" (Tracy, "Lighthousekeeping: *Bleak House* and the Crystal Palace," 29). The title of his article, moreover, not only suggests Dickens's interest in displays of popular science, but also puns on Esther's dual role as housekeeper and light/energy source.

51. Morus, *When Physics Became King*, 130.

52. Dickens, *Bleak House,* 39.

53. Marsden, *Watt's Perfect Engine: Steam and the Age of Invention,* 1, 7.

54. Nixon, " 'Lost in the vast worlds of wonder,' " 273.

55. Marsden, *Watt's Perfect Engine,* 11.

56. Dickens, *Bleak House,* 79.

57. Dickens, *Bleak House,* 365, 369.

58. This observation begins to suggest how my reading intersects with Chris Vanden Bossche's analysis of class discourse in *Bleak House.* It also suggests how the engineering valuation of work converges with middle-class depictions of "an idle, unproductive land-owning class" and the corresponding contention that its own productivity made the middle class more fit to run the nation ("Class Discourse," 14). As he explores the novel's representation of "class identities that are instable and discontinuous" (p. 27), Vanden Bossche interprets in terms of class many of the same things this chapter interprets in the terms of engineering: the energies of Mrs. Jellyby and Mrs. Pardiggle, the studied inactivity of Mr. Skimpole and Mr. Turveydrop, Esther's industry, and the limited effectiveness of individual action in the context of the oppressive social system depicted in the novel.

59. Harvey Sucksmith makes this observation as he discusses Dickens's "deliberate fusing of inventors, industrialists, Chartist workers at torch-light demonstrations, and the Wat Tyler image." He argues that this satire is "not an instance of *gross* caricature on Dickens' part," but instead captures the complexity of the social situation, because the name "Wat Tyler" refers to two revolutionary figures. The first of these, Wat the Tiler, was the leader of a medieval peasants' revolt, with whom Dickens sympathized. The second was a contemporary of Dickens, who adopted the alias "Wat Tyler" and who was among the leaders of the Chartist riots of 1848. ("Sir Leicester Dedlock, Wat Tyler, and the Chartists," 124, 119). Both *Bleak House* and *A Tale of Two Cities* suggest Dickens's ambivalence regarding Chartist agitation for political and social reform. Where he seems sympathetic with the plight of the working classes, he also reveals considerable anxiety about the violence that accompanied their protests.

60. Dickens, *Bleak House*, 79.

61. Dickens, *Bleak House*, 7.

62. Dickens, *Bleak House*, 149.

63. Dickens, *Bleak House*, 82, 253, 261.

64. Dickens, *Bleak House*, 364.

65. Dickens, *Bleak House*, 110–111.

66. Dickens, *Bleak House*, 12.

67. Dickens, *Bleak House*, 9.

68. Dickens, *Bleak House*, 82.

69. Dickens, *Bleak House*, 536, 541.

70. Dickens, *Bleak House*, 283.

71. Dickens, *Bleak House*, 10, 536, 541, 475.

72. I continue to use these terms, of course, in the senses defined by Latour. *Actants*, in particular, are distinguished from *actors* in his formulation, because though either may be human or nonhuman, the former are specifically those that are represented by some kind of spokesperson. See Latour, *Science in Action*, 84–89.

73. Latour, *Science in Action*, 129.

74. Latour, *Science in Action*, 130 (emphasis mine).

75. I am not the first to consider the connection between *Bleak House* and machinery. In comparing the novel to the Great Exhibition of 1851, displayed in the Crystal Palace and intended as a tribute to England's industrial and material progress, Robert Tracy understands *Bleak House* as "Dickens's Great Exhibition of 1852, reminding England that all was not well, that the efficient machines so proudly shown and viewed were not matched by a correspondingly efficient social machinery" ("Lighthousekeeping," 45).

76. It is for this quality in Dickens's work that George Levine dubs him "the great novelist of entanglement." For Levine, Dickens's novels dem-

onstrate "those very connections of interdependence and genealogy that characterize Darwin's tangled bank" (Levine, *Darwin and the Novelists*, 119). Levine is referring to the last paragraph of *The Origin of Species*, in which Darwin expresses wonder at the complexity and interconnectedness of life: "It is interesting to contemplate a tangled bank, clothed with many plants of many kinds, with birds singing on the bushes, with various insects flitting about, and with worms crawling through the damp earth, and to reflect that these elaborately constructed forms, so different from each other, and dependent upon each other in so complex a manner, have all been produced by laws acting around us." That Darwin elsewhere in *Origin* refers to the "whole machinery of life" may suggest the connections between the notion of entanglement and that of machination.

77. Latour, *Science in Action*, 129.

78. Dickens, *Bleak House,* 104, 815, 24.

79. Bert Hornback observes frequent and significant connections between Esther and the omniscient narrator, and concludes—in ways that are suggestive of Esther as source—that Esther is "in some way the author, the *authority* for both her own shy and supposedly self-effacing portion of the narrative and for that invasive, present-tense omniscience" (Hornback, "The Narrator of *Bleak House*," 11).

Chapter 6

1. From "Out of the Sun" (1958). Cited in Westfahl, *Science Fiction Quotations*, 216.

2. George Levine's treatment of *Little Dorrit*, Gillian Beer's chapter in *Open Fields* on " 'The Death of the Sun,' " Jude Nixon's " 'Death Blots Black Out': Thermodynamics and the Poetry of Gerard Manley Hopkins," and even Allen MacDuffie's "Irreversible Transformations: Robert Louis Stevenson's *Dr. Jekyll and Mr. Hyde* and Scottish Energy Science" focus on the fear and disappointment wrought by the second law of thermodynamics. As we have seen, Stephen Brush, who identifies opposing affective trends, one for each law of thermodynamics, nonetheless relegates each to its time period (*The Temperature of History*, 29). And even

Crosbie Smith, whose cultural history *The Science of Energy* recounts an extremely complex network of social forces and implications associated with the developing science, chooses for his cover an image rife with entropic decay.

3. Smith, *The Science of Energy*, 100–101.

4. Wise, "Time Discovered and Time Gendered in Victorian Science and Culture," 54.

5. I am thinking of the programming term *daemon* or *demon*, which in operating systems like Unix, refers to programs that process, sort, or otherwise manipulate information in the background rather than under the direct control of the user. The "gamer" remark refers to the now classic video game called "Maxwell's Maniac" (Microsoft 1992), in which the player separates red and blue "molecules" into two halves of a divided box.

6. Hayles, *Chaos Bound*, 32.

7. MacDuffie, "Irreversible Transformations: Robert Louis Stevenson's *Dr. Jekyll and Mr. Hyde* and Scottish Energy Science."

8. Iwan Morus identifies Wells's closing pages as "a graphic description of Thomson's universe, though the materialist Wells provided his book with a cheeky evolutionist narrative that would have left Thomson fuming" (*When Physics Became King*, 141). And Bruce Clarke opens *Energy Forms* with an account of what he calls—alluding to J. G. Ballard's 1964 short story collection—"Terminal Beach" (p. 1).

9. Wilde, *The Picture of Dorian Gray*, 21.

10. Wilde, *The Picture of Dorian Gray*, 110, 77.

11. Wilde, *The Picture of Dorian Gray*, 108, 104.

12. Wilde, *The Picture of Dorian Gray*, 105.

13. "As often as the images of linear and circular time were presented during the Victorian era, they were, just as often, gendered: masculine for history, feminine for repetition" (Wise, "Time Discovered and Time Gendered in Victorian Science and Culture," 39).

14. Wilde, *Dorian Gray*, 105.

15. The literary-minded who are further interested in chaos can't do better than to read Katherine Hayles, whose account in *Chaos Bound* I draw on here to distinguish the two branches (pp. 9–10).

16. This fairly well-known information is treated at length in Ed Cohen's chapter "Legislating the Norm: From 'Sodomy' to 'Gross Indecency'" in *Talk on the Wilde Side.*

17. For more on the sexuality of *Dracula*, see Phyllis Roth, John Allen Stevenson, Talia Schaffer, and Christopher Craft, "'Kiss Me with Those Red Lips.'" Steven D. Arata considers how *Dracula* reflects a variety of anxieties associated with British colonialism ("The Occidental Tourist: *Dracula* and the Anxiety of Reverse Colonization"), and Jennifer Wicke places these anxieties in conversation with those wrought by mass media and mass consumption ("Vampiric Typewriting: *Dracula* and Its Media"). Jennifer Fleissner's aptly named "Dictation Anxiety" considers *Dracula* as wrestling with the ambivalence wrought by the increasing presence of women in the clerical workforce ("Dictation Anxiety: The Stenographer's Stake in *Dracula*").

18. Stoker, *Dracula*, 215.

19. See my essay "Reproducing Empire: *Moreau* and Others."

20. Stoker, *Dracula*, 251.

21. Hayles, *Chaos Bound*, 40.

22. Bruce Clarke associates this balance with "the new relations forged at mid-twentieth century between information and entropy [which] highlighted not only the potential positivities of randomness and disorder, but also the crucial difference between environmental isolation and openness. Even given the assumption of a universal tendency toward maximal entropy, organisms maintain their local organizations through *operational* closure structurally coupled to environmental perturbation by energy and data sources, and modern electronic systems process energy through informatic feedback" ("From Thermodynamics to Virtuality," 28).

23. Wicke, "Vampiric Typewriting," 489.

24. Thomson, *Mathematical and Physical Papers*, 511, 514 (emphasis mine).

25. See chapter 4, note 78.

26. Stoker, *Dracula*, 186.

27. Stoker, *Dracula*, 325.

28. Stoker, *Dracula*, 133, 237, 202, 324.

29. Stoker, *Dracula*, 23, 199.

30. Stoker, *Dracula*, 310, 263.

31. Stoker, *Dracula*, 247.

32. Stoker, *Dracula*, 227.

33. Stoker, *Dracula*, 263.

34. Stoker, *Dracula*, 309–310.

35. Stoker, *Dracula*, 155.

36. Stoker, *Dracula*, 227–228. The convertibility of matter and energy seems to herald the coming of Einstein's $E = mc^2$, perhaps the world's most familiar equation, which suggests that matter may be transformed into an amount of energy proportional to its mass. The conversion factor, the speed of light or c squared, suggests that a little mass goes a long way. If only . . .

37. Much of my sense of this came from collaborating with Russel Kauffman on a paper titled "Poetry for Physicists: Rereading Maxwell," presented at the Society for Literature and Science in Buffalo, New York (Fall 2001).

38. Stoker, *Dracula*, 267.

39. Cited in Hayles, *Chaos Bound*, 42.

40. Schneider and Sagan, *Into the Cool*, 62.

41. G. J. Barker-Benfield reminds us of the sexual implications of such expenditure in nineteenth-century America, in particular, the threat that marriage and women represented to men's energy. This sexual economy is implicit the term *spend*, which, he reminds us, "meant to reach orgasm. It was the shortened form of the more refined term 'expenditure' " (*The Horrors of the Half-Known Life: Male Attitudes toward Women and Sexuality in Nineteenth-Century America*, 12).

42. In chemistry, the phrase *triple point* is used to refer to the state of temperature and pressure at which the solid, liquid, and gas phases all coexist in thermodynamic equilibrium.

43. Stoker, *Dracula*, 10, 14, 25.

44. Stoker, *Dracula*, 228.

45. Stoker, *Dracula*, 263.

46. Hayles, *Chaos Bound*, 150.

47. Stoker, *Dracula*, 10.

48. Hayles, *Chaos Bound*, 100.

49. Morus, *When Physics Became King*, 267–268.

50. Stoker, *Dracula*, 218.

51. Wise, "Time Discovered and Time Gendered in Victorian Science and Culture," 45.

52. Stoker, *Dracula*, 207.

53. Edward Clarke elaborated this position in *Sex in Education; or, A Fair Chance for the Girls* (1873), as did Henry Maudsley: "It is not a mere question of smaller and larger muscles, but of the energy and power of endurance of the nerve-force which drives the intellectual and muscular machinery; not a question of two bodies and minds that are in equal physical conditions, but of one body and mind capable of sustained and regular hard labour, and of another body and mind which for one quarter of each month during the best years of life is more or less sick and unfit for hard work" ("Sex in Mind and Education," *Fortnightly*

Review 15 (1874): 479; cited in Morus, *When Physics Became King*, 267). For more on this line of argument, see Cynthia Russett's chapter "The Machinery of the Body" in *Sexual Science*.

54. Stoker, *Dracula*, 326–327.

55. Hayles, *Chaos Bound*, 174.

56. Stoker, *Dracula*, 295, 307.

57. See, for example, my articles on Wells, Wilde, and Haggard ("Reproducing Empire: *Moreau* and Others"; "The Domination of *Dorian Gray*"; "Embracing the Corpse: Discursive Recycling in H. Rider Haggard's *She*"). In American literature, Mark Seltzer connects the problem of masculine (re)production in the late nineteenth-century naturalist novel, to gendered deployments of thermodynamics, wherein males embody the second law and females, the first (*Bodies and Machines*, 39).

58. Stoker, *Dracula*, 218.

59. Winthrop-Young, "Undead Networks: Information Processing and Media Boundary Conflicts in *Dracula*," 115.

60. Stoker, *Dracula*, 203. Mina refers to a technology for making multiple copies at once, but the term *manifold* also has a meaning in topology that may be of interest here. It describes a curved space whose global structure may differ considerably from its local structure. Locally, this space seems to follow the laws of Euclidean geometry (the geometry of ordinary experience, in which lines and curves behave as they would on a flat plane), but globally, it may exhibit more complex behaviors. Consider, for example, a large sphere. Small lines drawn on its surface will look straight, but relatively long lines will exhibit curvature. Indeed, a large enough triangle drawn on such a surface will have interior angles summing to more than 180 degrees.

61. Stoker, *Dracula*, 326.

62. Wicke, "Vampiric Typewriting," 476.

63. Hayles, *Chaos Bound*, 7.

64. Wicke, "Vampiric Typewriting," 471.

65. Thomas Richards identifies this connection in the very roots of Victorian thermodynamics: "Entropy was from the start primarily an index of information." The attempts that Victorians made to control entropy in turn "contributed to the formation of a mythology of state control over knowledge in which the dream of controlled entropy played a central role" (*The Imperial Archive: Knowledge and the Fantasy of Empire*, 80, 85).

66. Of course, this increase of women into the clerical workforce produces its own anxieties. Considering how *Dracula* reveals the relation between "information and entropy . . . to be a gendered one," Jennifer Fleissner identifies a "dictation anxiety" attached to "broader fears about promiscuous data" that come from wider (female) access to information, as well as to the threat to the liberal subject, a threat to the sense of self, wholeness, and autonomy posed by "dictated prose [that] opens a seemingly unresolvable gap between the self that writes and that self's writing" ("Dictation Anxiety," 430, 433). Fleissner builds on Jennifer Wicke's work, which considers the "rationalization . . . of the procedures of bureaucracy and business, the feminization of the clerical work force, [and] the standardization of mass business writing" as well as the text's engagement with mass culture more broadly ("Vampiric Typewriting," 471). Geoffrey Winthrop-Young observes that "too many goods and bodies were moving too fast; the need to control them prompted considerable bureaucratic differentiation and systemization, including the introduction of special-purpose human operators. . . . formal and technological innovations . . . crucial to *Dracula*" ("Undead Networks," 112). And Katherine Hayles reads Henry James's "In the Cage" (1898) as the attempt by the female telegraph operator to tap into the dream of information to escape an economy of scarcity in "Escape and Constraint: Three Fictions Dream of Moving from Energy to Information."

67. Hayles, *Chaos Bound*, 235.

68. Gagnier, "Evolution and Information, or Eroticism and Everyday Life, in *Dracula* and Late Victorian Aestheticism," 151.

69. Stoker, *Dracula*, 211.

70. Stoker, *Dracula*, 250.

71. Winthrop-Young, "Undead Networks," 125.

72. Hayles, *Chaos Bound*, 236. Finding in the gothic form of *Dracula* and others, a "dangerous desire for transcendence," which is itself "a desire for virtuality—a life without bodies or narrative constraint," Jules Law also identifies the similarities between the gothic and various information technologies: "The whole point of the gothic as narrative form is to demonstrate its superiority to rival genres and media (diaries, newspapers, shorthand, telegraph, phonograph, hypnosis, telepathy) in approaching the condition of virtuality without assuming the risks of negation thereby entailed" (Law, "Being There: Gothic Violence and Virtuality in *Frankenstein, Dracula,* and *Strange Days*," 976, 986–987).

73. Stoker, *Dracula*, 209–210.

74. This paragraph briefly summarizes what Morus discusses at length in his section "The Other World" (*When Physics Became King*, 174–190). The quotes are from pages 174, 180–181, 188.

75. Stewart and Lockyer, "Sun as a Type," 325–327.

76. Stewart and Lockyer, "Sun as a Type," 324. For more on this, see chapter 3.

77. Latour, *Science in Action*, 108.

78. I am not the first to be unable to resist this pun, in order to signal what Stephen D. Arata calls the "anxiety of reverse colonization." Salman Rushdie's variant of the phrase has become a familiar reference to postcolonial literature since the publication of *The Empire Writes Back: Theory and Practice in Post-Colonial Literature* (1989).

79. Latour, *Science in Action*, 117.

80. Stoker, *Dracula*, 288, 55. The worry that Mina might not translate her loyalties in this way, but rather ally herself with enemies, is a concern that Jennifer Fleissner identifies as attached to clerical workers and work in general. "Can a self," she asks, "denuded enough to pledge absolute loyalty to another party, to serve as its 'agent,' avoid taking up that same position everywhere it turns, and hence embodying absolute

disloyalty?" Mina must therefore be "seen to treat typewriting as an extension of her marital duties" ("Dictation Anxiety," 433, 428).

81. Stoker, *Dracula*, 314.

82. Hayles, *Chaos Bound*, 236.

83. Winthrop-Young, "Undead Networks," 114–115.

84. Wicke, "Vampiric Typewriting," 485.

85. Jenkins, *Space and the "March of Mind,"* 211.

Epilogue

1. "The Future," *The Power of the Sun*, DVD, written by John Perlin and David Kennard. As it happens, Walter Kohn's 1998 Nobel Prize for the development of density functional theory was in chemistry.

2. Virginia Woolf, *The Waves*. Throughout her work, Woolf wrestles with such concerns wrought by atomic and quantum as well as relativistic physics.

3. I am thinking of two research projects in particular. Inspired by photosynthesis in plants, Daniel Nocera's group at MIT is developing a catalytic process to split water into hydrogen and oxygen for use in fuel cells (see Trafton, "'Major Discovery' from MIT Primed to Unleash Solar Revolution: Scientists Mimic Essence of Plants' Energy Storage System"). At the University of Pennsylvania, Andrew Rappe's group is working to develop ferroelectric, semiconducting solar materials that will better use the whole of the visible spectrum and not just the ultraviolet part (see Goldstein, "Solar-Powered Cars May Not Be Just Science Fiction Anymore: Penn Receives Federal Grant for New Research").

Appendix

1. Cited in Campbell and Garnett, *Life*, 646–648.

References

Adams, James Eli. "Woman Red in Tooth and Claw: Nature and the Feminine in Tennyson and Darwin." *Victorian Studies* 33, no. 1 (Autumn 1989): 7–27. Also reprinted in Rebecca Stott, ed., *Tennyson* (New York: Longman, 1996).

"Age of the Sun and Earth." *Cornhill Magazine* 38 (July–December 1878): 321–341.

Arata, Stephen D. "The Occidental Tourist: *Dracula* and the Anxiety of Reverse Colonization." *Victorian Studies* 33, no. 4 (Summer 1990): 621–645.

Armstrong, Isobel. "Tennyson in the 1850s: From Geology to Pathology—*In Memoriam* (1850) to *Maude* (1855)." In Philip Collins, ed., *Tennyson: Seven Essays*, 102–140. New York: St. Martin's Press, 1992.

Ashcroft, Bill, Gareth Griffiths, and Hellen Tiffin, eds. *The Empire Writes Back: Theory and Practice in Post-Colonial Literature.* New York: Routledge, 1989.

Asimov, Isaac, and Jason A. Shulman, eds. *Isaac Asimov's Book of Science and Nature Quotations.* New York: Weidenfeld & Nicolson, 1988.

Bacon, Alan. "English Literature Becomes a University Subject: King's College, London as Pioneer." *Victorian Studies* 29, no. 4 (Summer 1986): 591–612.

Barker-Benfield, G. J. *The Horrors of the Half-Known Life: Male Attitudes toward Women and Sexuality in Nineteenth-Century America*. New York: Harper & Row, 1976.

Beer, Gillian. *Darwin's Plots: Evolutionary Narrative in Darwin, George Eliot, and Nineteenth-Century Fiction*. Cambridge: Cambridge University Press, 2000. (Orig. pub. Routledge 1983.)

Beer, Gillian. *Open Fields: Science in Cultural Encounter*. Oxford: Clarendon Press, 1996.

Beer, John. "Tennyson, Coleridge, and the Cambridge Apostles." In Philip Collins, ed., *Tennyson: Seven Essays*, 1–35. New York: St. Martin's Press, 1992.

Blake, William. "The Voice of the Devil." In William Blake, *The Marriage of Heaven and Hell*. 1790. http://www.gailgastfield.com/mhh/mhh.html.

Bodenheimer, Rosemarie. *Knowing Dickens*. Ithaca, NY: Cornell University Press, 2007.

Brush, Stephen. *The Temperature of History: Phases of Science and Culture in the Nineteenth Century*. New York: Burt Franklin & Co., 1978.

Buckland, Adelene. "'The Poetry of Science': Charles Dickens, Geology, and Visual and Material Culture in Victorian London." *Victorian Literature and Culture* 35, no. 2 (2007): 679–694.

Bulwer-Lytton, Edward. *The Coming Race*. Santa Barbara, CA: Woodbridge Press, 1979.

Bulwer-Lytton, Edward. *The Coming Race*. Edited with an introduction by Peter W. Sinnema. Peterborough, Ontario: Broadview Press, 2008.

Butts, Richard. "Languages of Description and Analogy in Victorian Science and Poetry." *English Studies in Canada* 11, no. 2 (June 1985): 193–213.

Campbell, Lewis, and William Garnett. *The Life of James Clerk Maxwell*. London: Macmillan, 1882. Reprint, New York: Johnson Reprint Corporation, 1969.

Cantor, Geoffrey, and Sally Shuttleworth, eds. *Science Serialized: Representations of the Sciences in Nineteenth-Century Periodicals.* Cambridge, MA: MIT Press, 2004.

Carnot, Sadi. "Reflections on the Motive Power of Fire." Trans. and ed. R. H. Thurston. In E. Mendoza, ed., *Reflections on the Motive Power of Fire and Other Papers on the Second Law of Thermodynamics*, 1–59. New York: Dover, 1960. Reprint, Gloucester, MA: Peter Smith, 1977.

Chapple, J. A. V. *Science and Literature in the Nineteenth Century.* London: Macmillan Education, 1986.

Chatterjee, Kalika Ranjan. *Studies in Tennyson as a Poet of Science.* New Delhi: S. Chand & Co (Pvt), 1974.

Clarke, Bruce. "Allegories of Victorian Thermodynamics." *Configurations* 4, no. 1 (Winter 1996): 67–90.

Clarke, Bruce. *Energy Forms: Allegory and Science in the Era of Classical Thermodynamics.* Ann Arbor: University of Michigan Press, 2001.

Clarke, Bruce. "From Thermodynamics to Virtuality." In Bruce Clarke and Linda Dalrymple Henderson, eds., *From Energy to Information: Representation in Science and Technology, Art, and Literature*, 17–33. Stanford, CA: Stanford University Press, 2002.

Clarke, Edward H. *Sex in Education; or A Fair Chance for the Girls.* Boston: James R. Osgood and Company, 1873. Reprint, New York: Arno Press, 1972.

Clausius, Rudolph. *The Mechanical Theory of Heat with Its Application to the Steam-Engine and to the Physical Properties of Bodies.* London: John Van Voorst, 1867.

Cohen, Ed. *Talk on the Wilde Side: Toward a Genealogy of a Discourse on Male Sexualities.* New York: Routledge, 1993.

Coleridge, Samuel Taylor. "Religious Musings: A Desultory Poem, Written on the Christmas Eve of 1794." http://www.usask.ca/english/barbauld/related_texts/religious_musings.html.

Coslett, Tess. *The Scientific Movement and Victorian Literature*. New York: St. Martin's Press, 1982.

Craft, Christopher. "'Descend, Touch, and Enter': Tennyson's Strange Manner of Address." *Genders* 1 (March 1988): 83–101.

Craft, Christopher. "'Kiss Me with Those Red Lips': Gender and Inversion in Bram Stoker's Dracula." *Representations* 8 (Fall 1984): 107–133.

Currie, Richard A. "Against the Feminine Stereotype: Dickens's Esther Summerson and Conduct Book Heroines." *Dickens Quarterly* 16, no. 1 (March 1999): 13–23.

Dale, Peter Allan. *In Pursuit of a Scientific Culture: Science, Art, and Society in the Victorian Age*. Madison: University of Wisconsin Press, 1989.

Darwin, Charles. *The Descent of Man and Selection in Relation to Sex*. Project Gutenberg, [1874] 2000. http://www.gutenberg.org/dirs/etext00/dscmn10.txt.

Darwin, Charles. *On the Origin of Species*. Project Gutenberg, [1st ed., 1859] 1998. http://www.gutenberg.org/dirs/etext98otoos11.txt.

Darwin, Charles. *The Origin of Species by Means of Natural Selection; or, the Preservation of Favoured Races in the Struggle for Life*. Project Gutenberg, [6th ed., 1872] 1999. http://www.gutenberg.org/dirs/etext99/otoos610.txt.

Dickens, Charles. *Bleak House*. New York: Bantam Books, 1992.

Dickens, Charles. *A Tale of Two Cities*. London: Everyman, 1994.

Duncan, David. *The Life and Letters of Herbert Spencer*. Vol. 1. New York: D. Appleton and Company, 1908.

Everitt, C. W. F. "Maxwell's Scientific Creativity." In Rutherford Aris, H. Ted Davis, and Roger H. Stuewer, eds., *Springs of Scientific Creativity: Essays on Founders of Modern Science*, 71–141. Minneapolis: University of Minnesota Press, 1983.

Fielding, K. J. "Dickens and Science?" *Dickens Quarterly* 13, no. 4 (December 1996): 200–216

Fielding, K. J., and Shu-Fang Lai. "Dickens, Science, and *The Poetry of Science*." *Dickensian* 91, no. 1 (Spring 1997): 5–10.

Flammarion, Camille. *Omega: The Last Days of the World*. Lincoln: University of Nebraska Press, [1894] 1999.

Fleissner, Jennifer. "Dictation Anxiety: The Stenographer's Stake in *Dracula*." *Nineteenth-Century Contexts* 22 (2000): 417–455.

Freedman, Carl. "London as Science Fiction: A Note on Some Images from Johnson, Blake, Wordsworth, Dickens, and Orwell." *Extrapolation* 43, no. 3 (Fall 2002): 251–262.

Fulweiler, Howard W. "Tennyson's *In Memoriam* and the Scientific Imagination." *Thought* 59 (September 1984): 296–318.

Gagnier, Regenia. "Evolution and Information, or Eroticism and Everyday Life, in *Dracula* and Late Victorian Aestheticism." In Regina Barreca, ed., *Sex and Death in Victorian Literature*. Bloomington: Indiana University Press, 1990.

Gibson, Walker. "Behind the Veil: A Distinction between Poetic and Scientific Language in Tennyson, Lyell, and Darwin." *Victorian Studies* 2, no. 1 (September 1958): 60–68.

Gleick, James. *Chaos: Making a New Science*. New York: Viking, 1987.

Gliserman, Susan. "Early Science Writers and Tennyson's 'In Memoriam': A Study in Cultural Exchange." *Victorian Studies* 18, no. 3 and no. 4 (March and June 1975): 277–308, 437–459.

Gold, Barri J. "The Consolation of Physics: Tennyson's Thermodynamic Solution." *PMLA: Publications of the Modern Language Association of America* 117, no. 3 (May 2002): 449–464.

Gold, Barri J. "The Domination of *Dorian Gray*." *Victorian Newsletter* 91 (Spring 1997): 27–30.

Gold, Barri J. "Embracing the Corpse: Discursive Recycling in H. Rider Haggard's *She*." *English Literature in Transition* 38, no. 3 (September 1995): 305–327.

Gold, Barri J. "Reproducing Empire: *Moreau* and Others." *Nineteenth-Century Studies* 14 (2000): 173–198.

Gold, Barri J., and Russel P. Kauffman. "Poetry for Physicists: Rereading Maxwell." Paper presented at the annual meeting of the Society for Literature and Science, Buffalo, NY, October 11–14, 2001.

Goldstein, Danny. "Solar-Powered Cars May Not Be Just Science Fiction Anymore: Penn Receives Federal Grant for New Research." *Daily Pennsylvanian*, June 21, 2007. http://media.www.dailypennsylvanian.com/media/storage/paper882/news/2007/06/21/News/SolarPowered.Cars.May.Not.Be.Just.Science.Fiction.Anymore-2917287.shtml.

Gore, Al. Speech on Renewable Energy, delivered July 17, 2008, Constitution Hall, Washington, DC. http://www.npr.org/templates/story/story.php?storyId=92638501.

Haight, Gordon S. "Dickens and Lewis on Spontaneous Combustion." *Nineteenth Century Fiction* 10, no. 1 (June 1955): 53–63.

Harman, P. M. *Energy, Force, and Matter: The Conceptual Development of Nineteenth-Century Physics*. Cambridge: Cambridge University Press, 1982.

Hayles, N. Katherine. *Chaos Bound: Orderly Disorder in Contemporary Literature and Science*. Ithaca, NY: Cornell University Press, 1990.

Hayles, N. Katherine, ed. *Chaos and Order: Complex Dynamics in Literature and Science*. Chicago: University of Chicago Press, 1991.

Hayles, N. Katherine. "Escape and Constraint: Three Fictions Dream of Moving from Energy to Information." In Bruce Clarke and Linda Dalrymple Henderson, eds., *From Energy to Information: Representation in Science and Technology, Art, and Literature*, 235–254. Stanford, CA: Stanford University Press, 2002.

Hornback, Bert G. "The Narrator of *Bleak House*." *Dickens Quarterly* 16, no. 1 (March 1999): 3–12.

Iltis, Carolyn. "Leibniz and the *Vis Viva* Controversy." *Isis* 62, no. 1 (Spring 1971): 21–35.

James, Edward, and Farah Mendelsohn, eds. *The Cambridge Companion to Science Fiction*. Cambridge: Cambridge University Press, 2003.

Jann, Rosemary. "Saved by Science? The Mixed Messages of Stoker's *Dracula*." *Texas Studies in Literature and Language* 2 (1989): 273–287.

Jenkins, Alice. *Space and the "March of Mind."* Oxford: Oxford University Press, 2007.

Joule, James Prescott. *The Scientific Papers*. London: Taylor and Francis, 1884.

Keats, John. *Letters of John Keats*. Selection edited by Robert Gittings. Oxford: Oxford University Press, 1970.

Korg, Jacob. "Astronomical Imagery in Victorian Poetry." In James Paradis and Thomas Postlewait, eds., *Victorian Science and Victorian Values: Literary Perspectives*, 137–148. New York: New York Academy of Sciences, 1981.

Kuhn, Thomas S. "Energy Conservation as an Example of Simultaneous Discovery." In Thomas S. Kuhn, *The Essential Tension: Selected Studies in Scientific Tradition and Change*. Chicago: University of Chicago Press, 1977.

Kuhn, Thomas S. *The Structure of Scientific Revolutions*. Chicago: University of Chicago Press, 1962.

Latour, Bruno. *Science in Action: How to Follow Scientists and Engineers through Society*. Cambridge, MA: Harvard University Press, 1987.

Latour, Bruno, and Steve Woolgar. *Laboratory Life: The Construction of Scientific Facts*. Princeton, NJ: Princeton University Press, 1986.

Law, Jules. "Being There: Gothic Violence and Virtuality in *Frankenstein*, *Dracula*, and *Strange Days*." *English Literary History* 73, no. 4 (2006): 975–996.

Le Guin, Ursula K. *The Left Hand of Darkness*. New York: Ace Books, 2000.

Leigh, Percival. "The Chemistry of a Candle." *Household Words*, August 3, 1850, 439–444.

Levine, George. *Darwin and the Novelists*. Chicago: University of Chicago Press, 1988.

Lewes, G. H. "Animal Heat." *Blackwood's Edinburgh Magazine* 84 (1858): 414–430.

Lightman, Bernard. "Scientists as Materialists in the Periodical Press: Tyndall's Belfast Address." In Geoffrey Cantor and Sally Shuttleworth, eds., *Science Serialized: Representations of the Sciences in Nineteenth-Century Periodicals*, 199–237. Cambridge, MA: MIT Press, 2004.

Lockyer, J. Norman. "What the Sun Is Made Of." *The Nineteenth Century*, July 1878, 75–87.

Lockyer, Sir Norman, and Winifred L. Lockyer. *Tennyson as a Student and Poet of Nature*. London: Macmillan, 1910.

MacDuffie, Allen. "Irreversible Transformations: Robert Louis Stevenson's *Dr. Jekyll and Mr. Hyde* and Scottish Energy Science." *Representations* 96 (Fall 2006): 1–20.

Maher, Brendan. "In This House We Obey the Laws of Thermodynamics: The Top Ten Science Moments in *The Simpsons*, as Chosen by *Nature*'s Editorial Staff." *Nature* 448 (July 26, 2007): 405.

Marsden, Ben. *Watt's Perfect Engine: Steam and the Age of Invention*. New York: Columbia University Press, 2002.

Mattes, Eleanor. *In Memoriam: The Way of a Soul*. New York: Exposition Press, 1951.

Maxwell, James Clerk. *The Scientific Papers of James Clerk Maxwell*. Vol. 2, ed. W. D. Niven. Cambridge: Cambridge University Press, 1890. Reprint, Mineola, NY: Dover, 2003.

Meyers, Greg. "Nineteenth-Century Popularizations of Thermodynamics and the Rhetoric of Social Prophecy." In Patrick Brantlinger, ed., *Energy and Entropy: Science and Culture in Victorian Britain*, 307–338. Bloomington: Indiana University Press, 1989.

Miller, D. A. *The Novel and the Police*. Berkeley: University of California Press, 1988.

Morus, Iwan Rhys. *When Physics Became King*. Chicago: University of Chicago Press, 2005.

Munro, J. "New Sources of Electric Power." *The Nineteenth Century*, December 1894, 919–931.

Nayder, Lillian. "Bulwer Lytton and the Imperial Gothic: Defending the Empire in *The Coming Race*." In Allan Conrad Christensen, ed., *The Subverting Vision of Bulwer Lytton: Bicentenary Reflections*, 212–221. Newark: University of Delaware Press, 2004.

Nixon, Jude. " 'Death Blots Black Out': Thermodynamics and the Poetry of Gerard Manley Hopkins." *Victorian Poetry* 40, no. 2 (Summer 2002): 131–155.

Nixon, Jude. " 'Lost in the vast worlds of wonder': Dickens and Science." *Dickens Studies Annual: Essays on Victorian Fiction* 35 (2005): 267–333.

Nunokawa, Jeff. "*In Memoriam* and the Extinction of the Homosexual." *English Literary History* 58 (1991): 427–438. Reprinted in Rebecca Stott, ed., *Tennyson* (New York: Longman, 1996).

O'Neill, Patricia. "Victorian Lucretius: Tennyson and the Problem of Scientific Romanticism." In J. B. Bullen, ed., *Writing and Victorianism*, 104–119. London: Longman, 1997.

Paradis, James, and Thomas Postlewait, eds. *Victorian Science and Victorian Values: Literary Perspectives*. New York: New York Academy of Sciences, 1981.

Paxton, Nancy L. *George Eliot and Herbert Spencer: Feminism, Evolutionism, and the Reconstruction of Gender*. Princeton, NJ: Princeton University Press, 1991.

Perlin, John, and David Kennard. *The Power of the Sun*. DVD. Produced by Walter Kohn, David Kennard, and Victoria Simpson. Mill Valley, CA: InCA Productions, 2005.

Peterfreund, Stuart. "The Re-Emergence of Energy in the Discourse of Literature and Science." *Annals of Scholarship* 4, no. 1 (1986): 22–53.

Rabinbach, Anson. *The Human Motor: Energy, Fatigue, and the Origins of Modernity.* Berkeley: University of California Press, 1990.

Rankine, William John Macquorn. *Miscellaneous Scientific Papers.* London: C. Griffin, 1881.

Richards, Thomas. *The Imperial Archive: Knowledge and the Fantasy of Empire.* New York: Verso, 1993.

Rosa, E. B. "The Human Body as Engine." *Popular Science Monthly* 57 (1900): 491–499.

Roth, Phyllis. "Suddenly Sexual Women in Bram Stoker's *Dracula.*" *Literature and Psychology* 27 (1977): 113–121.

Russett, Cynthia Eagle. *Sexual Science: The Victorian Construction of Womanhood.* Cambridge, MA: Harvard University Press, 1989.

Said, Edward. *Culture and Imperialism.* New York: Vintage Books, 1994.

Schaffer, Talia. "'A Wilde Desire Took Me': The Homoerotic History of *Dracula.*" *English Literary History* 61, no. 2 (1994): 381–425.

Schneider, Eric D., and Dorion Sagan. *Into the Cool: Energy Flow, Thermodynamics, and Life.* Chicago: University of Chicago Press, 2005.

Secord, James A. *Victorian Sensation: The Extraordinary Publication, Reception, and Secret Authorship of* Vestiges of the Natural History of Creation. Chicago: University of Chicago Press, 2000.

Seltzer, Mark. *Bodies and Machines.* New York: Routledge, 1992.

Serres, Michel. *Hermes: Literature, Science, and Philosophy.* Ed. Josue V. Harari and David F. Bell. Baltimore: Johns Hopkins University Press, 1982.

Siemens, William. "A New Theory of the Sun: The Conservation of Solar Energy." *The Nineteenth Century,* April 1882, 510–525.

Smith, Crosbie. *The Science of Energy: A Cultural History of Energy Physics in Victorian Britain.* Chicago: University of Chicago Press, 1998.

Snow, C. P. *The Two Cultures and the Scientific Revolution.* New York: Cambridge University Press, 1959.

Spencer, Herbert. *First Principles.* A reprint of the sixth and final edition, revised by the author in 1900. Honolulu: University Press of the Pacific, 2002.

Spencer, Herbert. *Principles of Biology.* New York, D. Appleton and Company, 1866–67.

Stevenson, John Allen. "A Vampire in the Mirror: The Sexuality of *Dracula.*" *PMLA: Publications of the Modern Language Association of America* 103, no. 2 (March 1988): 139–149.

Stewart, Balfour. *The Conservation of Energy.* New York: D. Appleton and Co., 1873. Citations are to the 1886 editon.

Stewart, Balfour, and Norman Lockyer. "The Sun as a Type of the Material Universe." Pts. 1 and 2. *Macmillan's Magazine,* July 1868, 246–257; August 1868, 319–327.

Stewart, B., and P. G. Tait. *The Unseen Universe or Physical Speculations on a Future State.* 9th ed. London: Macmillan and Co., 1890. Reprint, MT: Kessinger Publishing, 2003.

Stoker, Bram. *Dracula.* New York: Norton, 1997.

Stoppard, Tom. *Arcadia.* Boston: Faber and Faber, 1993.

Stott, Rebecca, ed. *Tennyson.* New York: Longman, 1996.

Sucksmith, Harvey Peter. "Sir Leicester Dedlock, Wat Tyler, and the Chartists: The Role of the Ironmaster in *Bleak House.*" *Dickens Studies Annual* 4 (1975): 113–131.

"The Sun's Surroundings and the Coming Eclipse." *Cornhill Magazine* 31 (January–June 1875): 297–315.

Tait, P(eter) G(uthrie). *Lectures on Some Recent Advances in Physical Science, with a Special Lecture on Force.* London: Macmillan and Co., 1885.

Tennyson, Alfred, Lord. *In Memoriam.* New York: Norton, 1973.

Thomson, William. "On the Age of the Sun's Heat." *Macmillan's Magazine*, March 1862, 388–393.

Thomson, William. "On the Dissipation of Energy." *Fortnightly Review*, March 1, 1892, 313–321.

Thomson, William. *Mathematical and Physical Papers*. Vol. 1. Cambridge: Cambridge University Press, 1882.

Tracy, Robert. "Lighthousekeeping: *Bleak House* and the Crystal Palace." *Dickens Studies Annual* 33 (2003): 25–53.

Trafton, Ann. "'Major Discovery' from MIT Primed to Unleash Solar Revolution: Scientists Mimic Essence of Plants' Energy Storage System." *MIT News*, July 31, 2008. http://web.mit.edu/newsoffice/2008/oxygen-0731.html.

Tyndall, John. *Heat: A Mode of Motion*. 1863. Citations are to the fourth edition. New York: D. Appleton and Company, 1873.

Tyndall, John. Presidential Address to the British Association of the Advancement of Science, Belfast, 1874. Victorian Web, www.victorianweb.org/science/science_texts/belfast.html.

Vanden Bossche, Chris R. "Class Discourse and Popular Agency in *Bleak House*." *Victorian Studies: An Interdisciplinary Journal of Social, Political, and Cultural Studies* 47, no. 1 (Autumn 2004): 7–31.

Viswas, Asha. *Tennyson's Romantic Heritage*. Meerut: Shalabh Prakashan, 1987.

Wells, H. G. *The Island of Doctor Moreau*. Rutland, VT: Tuttle, 1993.

Wells, H. G. *"The Time Machine" and "The War of the Worlds."* New York: Fawcett Premier, 1989.

Westfahl, Gary, ed. *Science Fiction Quotations: From the Inner Mind to the Outer Limits*. New Haven, CT: Yale University Press, 2005.

Whitworth, Michael. *Einstein's Wake: Relativity, Metaphor, and Modernist Literature*. Oxford: Oxford University Press, 2001.

Wicke, Jennifer. "Vampiric Typewriting: *Dracula* and Its Media." *English Literary History* 59, no. 2 (Summer 1992): 467–493.

Wilde, Oscar. *The Picture of Dorian Gray*. Chatham, Kent: Mackays of Chatham Place, 1994.

Wilkinson, Ann. "From Faraday to Judgment Day." *English Literary History* 34, no. 2 (June 1967): 225–247.

Williams, Rosalind. *Notes on the Underground: An Essay on Technology, Science, and the Imagination*. Cambridge, MA: MIT Press, 2008.

Winthrop-Young, Geoffrey. "Undead Networks: Information Processing and Media Boundary Conflicts in *Dracula*." In Donald Bruce and Anthony Purdy, eds., *Literature and Science*, 107–129. Atlanta: Rodopi, 1994.

Wise, M. Norton. "Time Discovered and Time Gendered in Victorian Science and Culture." In Bruce Clarke and Linda Dalrymple Henderson, eds., *From Energy to Information: Representation in Science and Technology, Art, and Literature*, 39–58. Stanford, CA: Stanford University Press, 2002.

Woolf, Virginia. *The Waves*. 1931. Project Gutenberg, 2002. http:// gutenberg.net.au/ebooks02/0201091.txt.

Wordsworth, William. "Ode on Intimations of Immortality from Recollections of Early Childhood." 1807. Bartleby.com. http://www.bartleby .com/145/ww331.html.

Young, Thomas. *A Course of Lectures on Natural Philosophy and the Mechanical Arts*. Vol. 1. 1807. Reprint, London: Johnson Reprint Corporation, 1971.

Zencey, Eric. "Entropy as Root Metaphor." In Joseph W. Slade and Judith Yaross Lee, eds., *Beyond the Two Cultures: Essays on Science, Technology, and Literature*, 185–200. Ames: Iowa State University Press, 1990.

Zwierlein, Anne-Julia. "The Evolution of Frogs and Philosophers: William Paley's *Natural Theology*, G. H. Lewes's *Studies in Animal Life* and Edward Bulwer-Lytton's *The Coming Race*." *Archiv für das Studium der Neueren Sprachen und Literaturen* 157, no. 2 (2005): 349–356.

Index

"Adonais" (Shelley), 50
Agnosticism, 139
Allen, Woody, 6
Ambiguity, 3–4, 115, 137–138,
    235, 240, 251
Anachronism, 35–41
Arcadia (Stoppard), 33
Artificial selection, 17
Ashton, Henry, 233
Asimov, Isaac, 6
Astronomy, 37, 131, 268n9

Barham, 115
Beer, Gillian, 28–29, 265n11
Biology
    electricity and, 88
    evolution and, 29 (*see also*
        Evolution)
    principles of energy and, 72
    Spencer and, 168
    Victorian physics and, 41–44, 81
    waste and, 65
Blackwood's, 27
Blake, William, 3–4, 26, 51

Bleak House (Dickens), 10, 18, 24,
    31, 86, 153
    Boythorn and, 194, 216
    Bucket and, 201
    Caddy and, 197–198, 206–207
    Carstone and, 195, 200–201
    Chancery and, 193–198, 201,
        204, 212, 215–217, 291n10
    Dedlock family and, 200–201,
        212–222, 227
    directing energy and, 198–202
    domestic engines and, 196, 205,
        207–212
    entropy and, 187–200, 204–205,
        209, 215, 218–221
    Gridley and, 195–196, 205
    Guppy and, 215–216, 218–219
    heat sinks and, 194–198,
        217–221
    Jarndyce & Jarndyce and,
        194–195, 201, 217
    Jellyby and, 197, 204–210, 217,
        236, 293n58
    Jo and, 198–202

Bleak House (Dickens) (*cont.*)
  Krook and, 196, 290n8
  maintaining distinction and,
    215–217
  making scientific facts and,
    189–192, 221–223
  Pardiggle and, 197, 204–207,
    217, 236, 293n58
  problematic prototypes and,
    204–208
  Rouncewells and, 212–214, 221
  Skimpole and, 195, 198–199,
    201–202
  spontaneous combustion and,
    190–191
  Summerson and, 193, 197, 201,
    206, 208–210, 218, 221–223,
    293n58, 295n79
  time of publishing, 188
  Tulkinghorn and, 217, 219–221,
    227–228, 230, 237
  Turveydrop and, 196–198
  wasting candles and, 193–194,
    206–207
  work and, 187, 189, 197–223
Boltzmann, Ludwig, 101
Boythorn, 194, 216
Breeding, 17
British Association, 27
British Society for Literature and
  Science, 22
Bucket, Inspector, 201
Bulwer-Lytton, Edward, 18, 24,
  31, 276n4
  engines and, 81
  evolution and, 76–79, 274n16

  Fiction Contest of, 75
  force and, 123
  grand unified theories and, 74–
    78, 81, 87, 92, 96–101, 107, 109
  literary expressions of, 75
  science fiction and, 30, 74–76,
    78, 89–90, 95–96, 100–101,
    116, 123, 147, 152, 154
Butterfly effect, 134
Butts, Richard, 271n56
Byron, 33, 46

Calculus, 120–121, 124–125
Caloric, 58–59, 74, 209, 258
Campbell, Alan, 233
Carnot, Sadi
  caloric and, 58–59, 209
  conservation of energy and, 38
  first law of thermodynamics
    and, 53
  force and, 128–131
  God and, 58
  heat engines and, 81, 141
  heat loss and, 60–62, 187
  Lockyer and, 146
  "On the Motive Power of Fire"
    and, 38
  steam engines and, 129–131, 149
  thermodynamics and, 13–14,
    26, 37–39, 47, 53, 58–61, 81,
    128–132, 141, 146, 149, 187,
    189, 209
  Thomson and, 38
Carnot cycle, 189
Carstone, Richard, 195, 200–201,
  205

Cartesian vortices, 117
Carton, Sydney, 152, 155
  energy of, 176–178
  as figure of positive
    transformation, 179–181
  novelistic order and, 187–188
  work of, 179–183
Chambers, Robert, 18
Chancery, 193–198, 201, 204,
  212, 215–217, 291n20
Chaos, 227, 245, 258, 292n50.
  See also Entropy
  butterfly effect and, 134
  Gleick and, 248
  information and, 248–252
  strange attractors and, 232–233,
    243–244
  women and, 248–252
"Chemical History of a Candle,
  The" (Faraday), 207
Christianity, 36, 91, 126–128
Clarke, Arthur C., 225
Clarke, Bruce, 28, 297n22
Clausius, Rudolph, 6, 38, 42,
  52–53, 124–125, 187
Coleridge, 50–51, 66, 115,
  268n18
Coming Race, The (Bulwer-
  Lytton), 276n4
  evolution and, 76–78
  force and, 123
  science fiction and, 30, 74–76,
    78, 89–90, 95–96, 100–101,
    116, 123, 147, 152, 154
Conservation of Energy, The
  (Stewart), 39

Continental-action-at-a-distance
  theories, 125
Coverly, Thomasina, 33
Cruncher, Jerry, 174–175
Cultural splitting, 21–24
Curie, Marie, 250, 254

"Darkness" (Byron), 33, 46
Darnay, Charles, 152, 165–166,
  173, 175–176, 179, 181
Darwin, Charles, 47, 139, 183.
  See also Evolution
  Bulwer-Lytton and, 78–79
  The Descent of Man and, 29–30
  On the Origin of Species and, 15,
    17, 36, 264n10, 285n14,
    294n76
  theology and, 26
  Thomson's opposition to,
    41–42
Darwin and the Novelists (Levine),
  28
Davy, Humphry, 18, 125
Dedlock family, 200–201, 212–
  222, 227
Defarge family, 159–161, 164,
  168–173
"Demon of the Second King"
  (Lem), 250
Descent of Man, The (Darwin),
  29–30
Dickens, Charles, 18, 24–25, 31.
  *See also Bleak House; A Tale of
  Two Cities*
  boiling water metaphor and,
    159–160

Dickens, Charles (*cont.*)
    cause and effect denial and, 163
    class oppression and, 162
    diffusion and, 223
    as engineer, 14
    engines and, 164, 264n3
    entropy and, 7–8, 151–156,
        161, 164, 169, 171–174, 178,
        181–182, 187–200, 204–205,
        209, 215, 218–221
    exposition of natural process
        and, 158–161
    insight into human behavior
        of, 285n14
    Lewes and, 191–192
    mechanics of the individual
        and, 169–173
    moral laws and, 162
    nature and, 152, 158–160,
        164–167, 180, 193, 211,
        216
    Newcomen steam pump and,
        189, 210–213, 222
    as novelist of entanglement,
        294n76
    science fiction and, 290n16
    social context and, 163–165
    spontaneous combustion and,
        190–192
    thermodynamics and, 7–8
    universal decay and, 192–193
    waste and, 173–181, 194, 198,
        201–206, 216
    wasting candles and, 193–194,
        206–207
    Watt and, 189, 211–214

Dickinson, 6
Diffusion, 9–10, 15, 36
    Bulwer-Lytton and, 87
    Dickens and, 223
    grand unified theories and,
        86–88, 105, 109–110
    Latour and, 269n19
    Spencer and, 87
    Tennyson and, 36, 55–56,
        62–67
Discovery, 39–41
Dissipation, 141, 226, 238–239,
    245
    conservation and, 25, 30, 34
        (*see also* Energy)
    death of sun and, 23–24, 28–29,
        44–49, 53–57, 108–109
    degradation and, 143–144
    Dickens and, 156–158, 199, 206,
        212, 220
    equilibrium and, 158 (*see also*
        Equilibrium)
    grand unified theories and, 77,
        83, 87, 100–101, 110
    heat sinks and, 9, 43, 53, 66,
        155, 177, 182, 194–196, 210–
        212, 220, 230, 233, 236, 239
    Maxwell-Boltzmann distribution
        and, 101–102
    progression and, 141
    Tennyson and, 38–45, 50, 62–67
    thermodynamic memory and,
        34 (*see also* Thermodynamics)
    Thomson and, 142–143, 237
    Wilde and, 227, 229–234, 243
Dissipative structures, 232–233

Dracula (Stoker)
  British colonialism and, 297n17
  condensers within, 237–241
  energy use of, 237–253,
    256–258
  engines of enlargement and,
    234–236
  entropy and, 24, 31, 227,
    234–253, 256–258, 297n17,
    301n66
  equilibrium and, 235–236
  gothic form of, 302n72
  Harker family and, 235–240,
    244–245, 248, 252–253
  sexuality and, 234–236, 245–
    252, 297n17
  sun and, 237–238
  Van Helsing and, 238, 246, 248,
    251–253, 257
Dr. Jekyll and Mr. Hyde
  (Stevenson), 227–229, 295n2
Duke of Perth, 233

Einstein, Albert, 3–4, 298n36
Electricity, 254
  as energy, 71, 146–147, 210
  Faraday and, 96
  as fluid, 27, 119
  imponderables and, 27
  Joule and, 59
  Maxwell and, 14–16, 46, 74,
    119–121, 125, 258
  Spencer and, 73, 146–147, 167
  Tennyson and, 37, 48–49, 54,
    66, 80
  Watt and, 211

Encyclopedia Brittanica, 27
Energy, 259
  Bleak House and, 187–223
  changing forms of, 7–9
  conservation of, 3, 6 (*see also*
    First law of thermodynamics)
  death of sun and, 23–24, 28–29,
    44–49, 53–57, 108–109
  defining, 4–7
  diffusion and, 9–10, 36, 55–56,
    62–63, 86–88, 105, 109–110
  dissipation and, 25, 30, 34 (*see
    also* Dissipation)
  Einstein and, 3–4, 298n36
  electricity and, 71, 146–147, 210
    (*see also* Electricity)
  in elegy, 51
  engine design and, 10–11
  entropy and, 6 (*see also*
    Entropy)
  equation of, 281
  fields and, 16–17
  force and, 93–100, 117 (*see also*
    Force)
  God and, 139–144
  grand unified theories and,
    71–72 (*see also* Grand unified
    theories)
  heat death and, 10, 36–37, 43,
    67, 104–108, 116, 123, 154–
    157, 163, 178, 182–183, 193–
    195, 215, 225
  heat sinks and, 9, 43, 53, 66,
    155, 177, 182, 194–196, 210–
    212, 220, 230, 233, 236, 239
  indestructibility of, 122

Energy (*cont.*)

  kinetic, 5, 16, 37, 61, 89–90,
    101, 114, 124, 132–133, 136,
    213–214, 280n11

  magnetism and, 5 (*see also*
    Magnetism)

  Maxwell-Boltzmann distribution
    and, 101–102

  meaning of, 4–5

  mechanical effect and, 11

  mechanics of the individual
    and, 169–173

  as metaphor, 4

  motion and, 71–72

  nuclear, 9

  other forms of, 5–6

  as persistence of force, 7

  potential, 5, 9, 53, 114, 132–
    133, 136, 166, 213–214,
    280n11

  progress of, 49–53

  protean quality of, 113–114, 167

  renewable, 265n15

  rest and, 105–110

  second law of thermodynamics
    and, 5–6

  social context and, 135–138,
    145–146

  spiritual correlates and, 49–53

  spontaneous combustion and,
    190–192

  states of, 180

  statistical mechanics and,
    100–104

  strange attractors and, 232–233,
    243–244

  Tennyson and, 37–43, 51, 60–
    61, 67

  thermal, 9–10, 89, 101, 239

  thermodynamics and, 5–6 (*see
    also* Thermodynamics)

  transforming language of,
    114–118

  unification from, 79–82

  as universal constant, 6

  Victorians and, 5

  Vril-ya and, 30, 74–76, 78, 89–
    90, 95–96, 100–101, 116, 123,
    147, 152, 154

  Watt and, 212–214

  work and, 5, 8–9, 11, 43 (*see
    also* Work)

Energy Forms (Clarke), 28

Engines

  Bleak House and, 187–223

  Carnot cycle and, 189 (*see also*
    Carnot, Sadi)

  Dickens and, 164, 264n3

  domestic, 196, 205, 207–212

  entropy and, 10–11, 187

  heat, 25, 86, 104, 108, 141, 203,
    210

  heat loss and, 60–62, 187

  Lockyer and, 147–148

  machination and, 222

  Newcomen steam pump and,
    189, 210–213, 222

  second law of thermodynamics
    and, 10–11

  social, 202–203

  Stewart and, 147–148

  Watt and, 211–214

Enlightenment, 129–130
Entropy, 31, 59, 133, 138,
    297n22, 301n65
    aging and, 229–233
    allegory and, 28
    Bleak House and, 187–189
    Bulwer-Lytton and, 98–100,
        109–112
    Carnot and, 38
    chaos and, 227, 232–233, 245,
        248–252, 258, 292n50
    chemical vs. physical approach
        to, 8
    Clausius and, 38–39, 42, 52–53,
        187
    death of sun and, 23–24, 28–29,
        44–49, 53–57, 108–109
    defining, 8–9
    Dickens and, 7–8, 151–156, 161,
        164, 169, 171–174, 178, 181–
        182, 187–200, 204–205, 209,
        215, 218–221
    Dracula and, 24, 31, 227, 234–
        253, 237–242, 245–247, 250,
        256–258, 297n17, 301n66
    engine design and, 10–11
    entropic individual and,
        227–231
    equilibrium and, 151–156, 161,
        164, 169, 171–174, 178, 181–
        182, 187
    force and, 98 (*see also* Force)
    increase of, 6
    information and, 248–252,
        301n66
    maximum, 6, 10, 42, 182

    mechanics of the individual
        and, 169–173
    poetry and, 7–9
    positive value of, 225
    progress of energy and, 49–53
    as root metaphor, 17
    second law of thermodynamics
        and, 10–11, 38, 42, 45
        (*see also* Second law of
        thermodynamics)
    Spencer and, 78, 83–84,
        109–110
    spontaneous combustion and,
        190–192
    statistical measurement and,
        226
    strange attractors and, 232–233,
        243–244
    A Tale of Two Cities and,
        151–183
    trope and, 52
    universal decay and, 192–193
    waste and, 48
    Wilde and, 227, 229–234, 243
Environmental issues, 23
Equilibrium, 30–31
    cessation of life and, 108
    death of sun and, 109
    diffusion and, 86–88
    disillusionment of, 25
    Dracula and, 235–236
    grand unified theories and,
        82–86
    moving, 110–112
    rest and, 105–110
    social, 82–86

Equilibrium (*cont.*)
  Spencer and, 151, 156–158
  A Tale of Two Cities and, 151–183
  two forms of, 157–158
  Vril-ya and, 82–86, 88, 104–105,
    110–112
Essay on Criticism (Pope), 123
Evolution, 25, 116, 229, 259
  affective relation with biology
    and, 41–44
  Bulwer-Lytton and, 76–79,
    274n16
  grand unified theories and, 76–
    79, 83–84, 105, 108–112
  On the Origin of Species and, 15,
    17, 36, 264n10, 285n14,
    294n76
  as ordering, 161
  physics and, 76–79
  rest and, 105
  social anxiety from, 36
  social Darwinism and, 31, 76
  Spencer and, 72–78, 156–157,
    163, 168, 182–183
  survival of the fittest and, 31,
    72, 112, 264n10
  Tennyson and, 36–37, 41–42,
    45–49, 62–66, 78
  as term of progress, 41
  Thomson's opposition to, 41–42
  value of organs and, 30
  waste and, 29–30

Faraday, Michael, 16, 80, 95–96,
    207
Fermi, Enrico, 21
Feynman, Richard, 13

Field theory, 16–17
First law of thermodynamics, 6,
    208, 239, 254
  Carnot and, 53
  Clausius and, 42
  God and, 139–144, 178, 187
  Newton and, 48
  progress of energy and, 49–53
  religion and, 187
  Tait and, 122, 178
  Tennyson and, 30, 42, 44, 48–
    49, 53–57, 64–66
  Thomson and, 38
First Principles (Spencer), 30,
    73–78, 91, 154, 156, 168
Flammarion, Camille, 46, 193,
    225
Force
  applied over distance, 203
  basic four of universe, 71–72
  Carnot and, 128–131
  Cartesian vortices and, 117
  definition of, 96
  division and, 93–100
  equation of, 281
  equilibrium and, 83–85 (*see also*
    Equilibrium)
  equivalents of power and, 96
  Faraday on, 95–96
  free will and, 134, 141,
    146–147
  God and, 139–144
  grand unified theories and,
    71–72, 79–82, 93–100
  gravity, 16, 133, 137–138, 142,
    155, 170–171, 216
  Joule and, 52, 117, 139–140

living, 52, 61, 88–93 (*see also* Kinetic energy)
Lockyer and, 131–139
Maxwell and, 95–96, 113–128
mechanical effect, 203
as mere vector, 115–121
Nature and, 158–161
Newton and, 47–48, 98, 117–118, 121–125
persistence of, 7, 73–74, 77–78, 81–84, 94–95, 111, 156–157, 167–169, 183
progress of energy and, 52
scale and, 133–138
social context of, 97–100
as space variation of energy, 121–122
Spencer and, 73–78, 91, 154, 156, 168
Stewart and, 125–128, 131–139
sun and, 131–133
Tait and, 95, 114–128
Tennyson and, 48
Thomson and, 139–140
Fortnightly Review, 27
Free will, 134, 141, 146–147, 165, 180, 255
French Reign of Terror, 98
French Revolution, 152–153

Galvanism, 80, 96
Gleick, James, 248, 283n51
God, 6, 15
Carnot and, 58
conservation of energy and, 139–144

creation and, 255–256, 258, 272n85
faith and, 51, 66
first law of thermodynamics and, 139–144, 178, 187
goodness of, 61
Joule and, 58, 61, 89, 139–142
Lockyer and, 141, 144, 255–256
as love, 58
monotheism and, 50, 66–67, 71
Nature and, 6, 15, 46, 49–50, 56, 63–64
perfection and, 15
politics and, 139–144
sovereign will of, 61
Stewart and, 141, 144, 255–256
Tennyson and, 46, 49–51, 56–58, 61–67
Thomson and, 58, 140–141
Tyndall and, 142
Grand unified theories, 25, 30, 67, 268n18
Bulwer-Lytton and, 74–78, 81, 87, 92, 96–101, 107, 109
diffusion and, 86–88, 105, 109–110
energy unification and, 79–82
equilibrium and, 82–86, 110–112
evolution and, 76–79, 83–84, 105, 108–112
force divisions and, 93–100
four basic forces and, 71–72
James Thomson and, 74, 81
Joule and, 95
living force and, 88–93

Grand unified theories (*cont.*)
  Lockyer and, 133–134
  molecular motion and, 72–73
  as One Great Law of science, 72
  positivism and, 72
  problem with rest and, 105–110
  Spencer and, 72–87, 91–99, 105,
    108–111
  statistical variation and,
    100–104
  Stewart and, 133–134
  Tait and, 71–72, 90–91, 95
  unified field theory and,
    275n23
  William Thomson and, 71–74,
    90–91, 95, 100
Gravity, 16, 133, 137–138, 142,
  155, 170–171, 216
Greenhouse gases, 23
Gridley, 195–196, 205
Guppy, 215–216, 218–219
Gwendolen, Lady, 233

Hallam, Arthur Henry, 34, 53,
  56, 62–63, 66
Hallward, Basil, 233
Harker family, 235–240, 244–245,
  248, 252–253
Hayles, N. Katherine, 225, 248,
  267n4, 302n72
Heat. *See also* Thermodynamics
  affective relation with biology
    and, 41–44
  caloric and, 58–59, 74, 209, 258
  Carnot cycle and, 189
  contradictory uses of, 58

death of sun and, 44–49, 53–57
definition of, 43
diffusion and, 9–10
engine design and, 10–11
fluid dynamics and, 160–161
heat death, 10, 36–37, 43, 67,
  104–108, 116, 123, 154–157,
  163, 178, 182–183, 193–195,
  215, 225
heat sinks, 9, 43, 53, 66, 155,
  177, 182, 194–196, 210–212,
  220, 230, 233, 236, 239
imponderables and, 27
loss of, 60–62
paradox of, 9–10, 25, 41–44
as thermal energy transfer, 10,
  89
*Heat Considered as a Mode of
  Motion* (Tyndall), 39, 88
Heat engines, 25, 86, 104, 108,
  141, 203, 210
Hetty, 233
Hood, 115
Hook, 115
Household Words magazine, 7,
  263n10
Humanists, 260
Huxley, Thomas, 91, 139

Imponderables, 27
Impressionism, 13
In Memoriam (Tennyson), 15–16,
  24
  anachronism in, 35–41
  consolation of physics and,
    57–62

death of sun and, 53–57
diffusion model and, 36
as elegy, 34
energy physics and, 37–43, 51,
  60–61, 67
evolution and, 36–37, 41–42,
  45–49, 62–66, 78
Hallam and, 34, 53, 56, 62–63,
  66
progress of energy and, 49–53
Romanticism and, 35, 37, 40,
  46–56, 66
second law of thermodynamics
  and, 41–56, 64–66
thermodynamic memory and,
  34–35
Interdisciplinarity, 19, 22
Interdisciplinary Nineteenth-
  Century Studies
  community, 22
International Centre for
  Theoretical Physics, 3
Irony, 3–4, 6, 99, 116, 162, 177

Jarndyce & Jarndyce, 194–195,
  201, 217
Jefferson, Thomas, 151
Jellyby family, 197–198, 204–210,
  217, 236, 293n58
Jo, 198–202
Joule, James Prescott, 16, 26,
  38–39, 125
electricity and, 59
entropy and, 187
force and, 52, 95, 126, 139–140
God and, 58, 61, 89, 139–142

grand unified theories and, 95
heat loss and, 60–62
initial resistance to
  thermodynamic theory and, 59
Nature and, 59–61

Kant Immanuel, 39, 268n18
Keats, John, 8
Kelvin temperature scale, 38
Kinetic energy, 213–214
Bulwer-Lytton and, 89–90
Joule and, 16, 61
Maxwell-Boltzmann distribution
  and, 101–102
potential energy and, 114,
  132–133, 136
social metaphor and, 5
Tait and, 124
Tyndall and, 280n11
Young and, 37
Kohn, Walter, 259
Krook, 196, 290n8
Kuhn, Thomas, 59

Lagrange, Joseph, 13–14
Laplace, Pierre-Simon, 39, 45, 53,
  268n18
Latour, Bruno, 21, 294n72
black box of, 257
diffusion model of, 36, 269n19
discovery and, 130
fact building and, 256
energy principles and, 152
machination and, 222
object redefinition and, 188
scientific article and, 113–114

Left Hand of Darkness, The
  (Le Guin), 17
Le Guin, Ursula, 17
Leibnitz, Wilhelm Gottfried,
  124–125, 282n31
Lem, Stanislaw, 250
Levine, George, 28
Lews, George Henry, 191–192
Light
  death of sun and, 23–24, 28–29,
    44–49, 53–57, 108–109
  first law of thermodynamics
    and, 53–57
  fluid metaphor for, 16–18
  imponderables and, 27
  Vril-ya and, 88
Lines of force, 16
Literature. *See also specific author*
  ambiguity and, 3–4, 115,
    137–138, 235, 240, 251
  cultural splitting and, 21–24
  discovery and, 39–41
  interdisciplinarity and, 19
  irony and, 3–4, 6, 99, 116, 162,
    177
  lunatics and, 33–34, 44, 193
  metaphor and, 4–5 (*see also*
    Metaphor)
  "Newton's sleep" and, 26
  nineteenth-century physics and,
    3–4
  productive conversation and,
    16–21
  puns and, 23, 115–116, 124, 128,
    214, 280n6, 292n50, 302n78
  Romanticism and, 26, 35, 37,
    40, 46–56, 66, 271n56

root metaphors and, 17–19
science and, 14–21
transformation in
  thermodynamics and, 114–118
tropes and, 40–41, 152,
  159–161
Lockyer, Norman, 26, 37, 214
  Carnot and, 146
  class relations and, 137–139
  dissipation of energy and, 143
  force and, 133–139
  God and, 141, 144, 255–256
  grand unified theories and,
    133–134
  second law of thermodynamics
    and, 145
  steam engine and, 147–148
  Stewart and, 131–133, 144
Lodge, Oliver, 254–255
Lorry family, 170–175, 188,
  287n52
Lunatics, 33–34, 44, 193
Lyell, Charles, 18

MacDuffie, Allen, 228
Macmillan's Magazine, 44
Magnetism, 254
  atmospheric, 88
  Bulwer-Lytton and, 80
  electromagnetism and, 71
  energy and, 5, 27
  Faraday and, 88, 96
  imponderables and, 27
  Maxwell and, 14–16, 74, 119,
    121, 258
  Spencer and, 167
  Stewart and, 90

Manette family, 155, 170–176,
    179–181
Mathematics, 13, 26, 171, 241,
    280n6
  calculus, 120–121, 124–125
  energy and, 4, 101–102
  entropy and, 8
  force and, 115, 118–121
  physical world and, 158
  statistical mechanics and, 226
  Thomson and, 74
Maxwell, James Clerk, 26, 91,
    253
  conundrum of, 226–227
  fluid metaphor of, 16–18
  electricity and, 14–16, 46, 74,
    119–121, 125, 258
  force and, 95–96, 113–128
  grand unified theories and, 74,
    95
  infinite continuity and, 66
  irreversible processes and, 163
  magnetism and, 14–16, 74, 119,
    121, 258
  poetry and, 16, 26–27, 113–128,
    261–262
  poetry of, 27, 261–262
  puns and, 115–116
  sorting demon of, 248–249
  statistical mechanics and, 101
  Tait and, 114–128, 242, 261–262
  unified field theory and,
    275n23
  Vis's and, 124–125
Maxwell-Boltzmann distribution,
    101–102
Mayer, J. R., 125

Mechanical effect, 11, 203
Mechanics, 13–14
  heat loss and, 60–62
  of the individual, 169–173
  Newtonian, 47
  statistical, 100–104
Menette, Doctor, 155
Metaphor, 27, 88–89, 258
  boiling water and, 159–160
  candles and, 193–194, 206–207
  continual, 94
  Dickens and, 159, 161, 178,
    193–194, 206–207, 213, 219,
    222
  electricity and, 119
  energy fields and, 4–5, 16–17
  entropy and, 17
  force and, 113–114, 118–119,
    126–134, 138, 145–149
  free will and, 134, 141, 146–147
  grand unified theories and,
    106–107, 110
  imponderables and, 27
  light as fluid, 16–18
  perpetual, 97
  root, 17–19
  scale and, 133–138
  science fiction and, 16–17, 89
  tendencies abroad to change
    and, 145–146
  Tennyson and, 47–53, 57, 61
  tropes and, 40–41, 152,
    159–161
Mill, John Stuart, 84–85, 95
Monroe Doctrine, 235
Monseigneur, 162–165, 168–171,
    188

Motion
 dissipation of, 156
 fluid friction and, 59
 gravity and, 167
 heat and, 39
 kinetic energy and, 5 (*see also*
  Kinetic energy)
 molecular, 9, 61, 157
 Newton and, 172
 perpetual, 162–165, 170, 176,
  181–182, 187, 228
 potential energy and, 166
 random, 9, 195
 Victorian concept of, 71–73, 77,
  83, 87–92, 119–120
Muhlenberg College, 22

Napoléon, 129
Naturalism, 72, 76–77, 91, 94–95,
 142, 273n3
Nature, 6, 17
 Dickens and, 152, 158–160,
  164–167, 180, 193, 211, 216
 forces of, 158–161
 gender of, 46–47, 64
 God and, 6, 15, 46, 49–50, 56,
  63–64
 Joule and, 59–61
 second law of thermodynamics
  and, 10
 Tennyson and, 30, 34, 36, 40,
  45–46, 49–50, 56, 59, 63–66
 Thomson and, 62
Nature journal, 27
Newcomen steam pump, 189,
 210–213, 222

Newton, Isaac, 4, 14, 16, 142
 Cartesian vortices and, 117
 energy and, 50
 force and, 47–48, 98, 117–118,
  121–125
 laws of motion and, 172
 Leibnitz and, 125
 perpetual motion and, 162–163
 Romanticism and, 37
 "Newton's sleep," 26
Nocera, Daniel, 303n3
North British School, 94–95, 122,
 126–127, 142, 273n3

*Omega: The Last Days of the
 World* (Flammarion), 46
"On a Universal Tendency in
 Nature to the Dissipation of
 Mechanical Energy"
 (Thomson), 62, 237
"On Collision" (Young), 37
"On Force" (Tyndall), 95
"On Matter, Living Force and
 Heat" (Joule), 58
"On the Conservation of Energy"
 (Maxwell), 95–96
"On the Dynamical Theory of
 Heat" (Thomson), 38
"On the Motive Power of Fire"
 (Carnot), 38
"On the Motive Power of Heat"
 (Carnot), 13, 129
On the Origin of Species (Darwin),
 15, 17, 36, 264n10, 285n14,
 294n76
Open Fields (Beer), 28–29

Pardiggle, 197, 204–207, 217, 236, 293n58

Perpetual motion, 162–165, 170, 176, 181–182, 187, 228

Physics, 259–260
 affective relation with biology and, 41–44
 closed systems and, 8
 disinterest in, 22–23
 energy and, 4–6, 49–53 (*see also* Energy)
 entropy and, 8, 180 (*see also* Entropy)
 evolution and, 76–79
 grand unified theories and, 25, 30 (*see also* Grand unified theories)
 imponderables and, 27
 metaphor and, 17–18 (*see also* Metaphor)
 naturalism and, 72, 76–77, 91, 94–95, 142, 273n3
 North British School and, 94–95, 122, 126–127, 142, 273n3
 poetry for, 3–11
 scientific affect and, 15, 24, 29, 41–44, 49, 62, 64, 155
 Tennyson and, 37–43, 51, 60–61, 67
 thermodynamics and, 7–8 (*see also* Thermodynamics)
 word use and, 3–4
Picture of Dorian Gray, The (Wilde)
 disorder in, 233–234
 dissipation and, 227, 229–234, 243

PMLA, 20

Poetry
 discovery and, 39–41
 energy and, 4–5
 entropy and, 7–9
 lunatics and, 33–34, 44, 193
 Maxwell and, 16, 26–27, 113–128, 261–262
 physics and, 3–11
 puns and, 115–116
 Romanticism and, 26, 35 (*see also* Romanticism)

Politics, 277n54
 God and, 139–144
 Monroe Doctrine and, 235
 Spencer and, 163
 thermodynamics and, 79–82, 85–86, 93, 96–100, 116
 Vril-ya and, 79–82, 85–86, 93, 96–100

Pope, 123

Positivism, 49, 72

Potential energy
 force and, 5, 9, 53, 114, 136, 166, 213–214, 280n11
 kinetic energy and, 114, 132–133, 136
 Maxwell and, 114
 sun and, 132–133
 Tennyson and, 53

Presbyterianism, 72, 140

Principia (Newton), 50

Principle of Continuity, 276n51

"Problem in Dynamics, A"
(Maxwell), 120–121
Progressivism, 41–42, 76, 225–226
Puns, 23, 115–116, 124, 128, 214,
280n6, 292n50, 302n78

Rappe, Andrew M., 259, 303n3
"Reflections from Various
Surfaces" (Maxwell), 16
*Réflexions sur la Puissance Motrice
du Feu* (Carnot), 129
Religion, 25, 127, 178, 273n3
agnosticism and, 139
Christianity and, 36, 91,
126–128
first law of thermodynamics
and, 187
grand unified theories and, 71,
82, 89
naturalism and, 72, 76–77, 91,
94–95, 142, 273n3
Presbyterianism and, 72, 140
Tennyson and, 40, 51, 66
"Religious Musings" (Coleridge),
50
"Report on Tait's Lecture on
Force" (Maxwell), 115, 121–
123, 261–262
Rest, 105–110, 155
Roentgen rays, 254–255
Romanticism
Blake and, 26
Newtonian knowledge and, 37
positivism and, 49, 72
progress of energy and, 49–53

Tennyson and, 35, 37, 40,
46–56, 66, 271n56
Turner and, 13
Rouncewell family, 212–214, 221
Royal Society of Edinburgh, 5
Ruskin, John, 52

Science
affective effects of, 15, 24, 29,
41–44, 49, 62, 64, 155
allegory and, 274n8
art and, 14
cultural splitting and, 21–24
death of sun and, 23–24, 28–29,
44–49, 53–57, 108–109
diffusion model and, 36
discovery and, 39–41
grand unified theories and, 25,
30 (*see also* Grand unified
theories)
historical spiritual modes of,
127–128
illiteracy of scientists and,
22–23
interdisciplinarity and, 19
literature and, 14–15
Maxwell's conundrum and,
226–227
metaphor and, 16 (*see also*
Metaphor)
naturalism and, 72, 76–77, 91,
94–95, 142, 273n3
Newtonian mechanics and, 47
"Newton's sleep" and, 26
physics and, 22 (*see also* Physics)

Principle of Continuity and, 276n51
productive conversation and, 16–21
root metaphors and, 17–19
univocality and, 280
Science fiction
Bulwer-Lytton and, 30, 74–76, 78, 89–90, 95–96, 100–101, 116, 123, 147, 152, 154
Carnot and, 129
Le Guin and, 17
metaphor and, 89
Science in Action (Latour), 113–114
Scientific literacy, 23
Second law of thermodynamics, 255, 269n35, 291n20, 295n2
death of sun and, 23–24, 28–29, 44–49, 53–57, 108–109
as degenerative process, 274n16
Dickens and, 153, 161–163, 169, 171, 194
dissipation and, 226–228 (*see also* Dissipation)
force and, 125, 133, 139–145
grand unified theories and, 72, 83, 99
Lockyer and, 145
as male, 300n57
progress of energy and, 49–53
Stewart and, 145
Stoker and, 235, 242–243
Tennyson and, 41–50, 55–56, 64–66

transfiguration and, 140
waste and, 44–49 (*see also* Waste)
Wilde and, 231–234, 243
Serres, Michel, 13–14, 151, 161
Seward, Dr., 249, 252
Sex, 259, 292n47
behavioral studies and, 269n35
Dickens and, 292n47
Dracula and, 234–236, 245–252, 297n17
fantasy and, 248–252
Victorians and, 25, 62–63, 75, 248–252, 299nn41,53
Shelley, Mary, 27, 50
Simpson family, 187
Singleton, Adrian, 233
Skimpole, 195, 198–202
Snow, C. P., 22–23
Social Darwinism, 31, 76
Society for Literature, Science, and the Arts, 22
Spencer, Herbert, 7, 18, 24, 30–31
anxiety of indistinction and, 163–164
counterentropic transformations and, 182–183
diffusion and, 87
equilibrium and, 83–86, 151, 156–158
evolution and, 72–78, 156–157, 163, 168, 182–183
expansive force and, 84
grand unified theories and, 72–87, 91–99, 105, 108–111

Spencer, Herbert (*cont.*)
  living force and, 91–92
  mechanics of the individual
    and, 170–172
  molecular motion and, 72–73
  moral laws and, 162
  persistence of force and, 94–95,
    167–169
  social context and, 163–164
  "survival of the fittest" phrase
    and, 31, 72
  Tyndall and, 151, 157–158
  unification and, 81–82
  Unknown Reality of, 92
  waste and, 78–79
Spontaneous combustion,
  190–192
Statistical mechanics, 100–104
Steam engines. *See* Engines
Stevenson, Robert Louis, 227–
  229, 295n2
Stewart, Balfour, 26, 39, 90,
  214
  class relations and, 137–139
  dissipation of energy and, 143
  force and, 131–139
  God and, 141, 144, 255–256
  grand unified theories and,
    133–134
  Lockyer and, 131–133, 255–256
  Principle of Continuity and,
    276n51
  second law of thermodynamics
    and, 145
  steam engine and, 147–148
  Tait and, 125–128

  universe as machine and, 187
  unseen universe and, 90–91
Stoker, Bram, 16, 24–26, 31, 235.
  See also Dracula (Stoker)
Stoppard, Tom, 33
Strange attractors, 232–233,
  243–244
Summerson, Esther, 293n58,
  295n79
  Bucket and, 201
  candle metaphor and, 206–207
  engine metaphor and, 208–210
  Lady Dedlock and, 218,
    221–223
  as narrator, 193
  Turveydrop and, 197
Sun, 260, 278n82
  death of, 23–24, 28–29, 44–49,
    53–57, 108–109
  Dracula and, 237–238
  first law of thermodynamics
    and, 53–57
  gravity and, 133
  Laplace's Nebular Hypothesis
    and, 45, 53
  Lockyer and, 131–133
  potential energy and, 132–133
  second law of thermodynamics
    and, 44–49, 53–57
  Thomson and, 45, 131–132
  Wells and, 45–46
"Sun as a Type of the Material
  Universe, The" (Stewart and
  Lockyer), 132–133
Survival of the fittest, 31, 72,
  112, 264n10

Tait, P. G., 26, 148
force and, 95, 114–128
grand unified theories and,
    71–72, 90–91, 95
Maxwell and, 114–128, 242,
    261–262
Principle of Continuity and,
    276n51
puns and, 115
scientific univocality and, 280
Stewart and, 125–128
unseen universe and, 90–91
Tale of Two Cities, A (Dickens),
    24, 31, 151
Carton and, 152, 155, 176–183,
    187–188
counterentropic transformations
    and, 180–183
Darnay and, 152, 165–166, 173,
    175–176, 179, 181
Defarge family and, 159–161,
    164, 168–173
dissolution and, 162–165
fluid dynamics and, 160–161
forces of nature and, 158–162
freedom and, 165–166
French Revolution and, 152–153
heat and, 160–161
Lorry family and, 170–175, 188,
    287n52
Manette family and, 155, 170–
    176, 179–181
mechanics of the individual
    and, 169–173
Monseigneur and, 162–165,
    168–171, 188

perpetual motion and, 162–163
persistence of force and,
    167–169
restored energy and, 155–156
social stability and, 154
Spencer's equilibria and,
    156–158
stability as fantasy in, 154–155
Stryver and, 176–177
tacit knowledge in, 152–153
ultimate equilibration and, 154
Tennyson, Alfred, Lord, 14, 16,
    18, 20, 24–25
consolation of physics and,
    57–62
death of sun and, 44–49, 53–57
diffusion and, 36, 55–56, 62–67
evolution and, 36–37, 41–42,
    45–49, 62–66, 78
first law of thermodynamics
    and, 30, 42, 44, 48–49, 53–57,
    64–66
force and, 48
God and, 46, 49–51, 56–58,
    61–67
Hallam and, 34, 53, 56, 62–63, 66
interest of in science, 37
Nature and, 30, 34, 36, 40, 45–
    46, 49–50, 56, 59, 63–66
as Poet Laureate, 15, 34
progress of energy and, 49–53
Romanticism and, 35, 37, 40,
    46–56, 66, 271n56
scales of phenomena and, 47
scientific affect and, 41–42, 49,
    62, 64

Tennyson, Alfred, Lord (*cont.*)
 second law of thermodynamics
  and, 30
 spiritual correlates and, 49–53
 systems applications and, 47–49,
  55–56
 thermodynamic memory and,
  34–35
 use of anachronism and, 35–41
 use of metaphor and, 47–53, 57,
  61
 waste and, 39, 43–48, 57, 61–66,
  78–79
Theology, 26, 37. *See also*
 Religion
Theory of Heat (Maxwell), 242
Thermal energy, 9–10, 89, 101,
 239
Thermodynamic memory, 34–35
Thermodynamics, 259
 affective relation with biology
  and, 41–44
 caloric and, 58–59, 74, 209, 258
 Carnot cycle and, 189 (*see also*
  Carnot, Sadi)
 closed systems and, 8, 42–43
 death of sun and, 23–24, 28–29,
  44–49, 53–57, 108–109
 dynamics of, 42–43
 engine design and, 10–11 (*see
  also* Engines)
 entropy and, 6 (*see also* Entropy)
 heat death and, 10, 36–37, 43,
  67, 104–108, 116, 123, 154–
  157, 163, 178, 182–183, 193–
  195, 215, 225

heat sinks and, 9, 43, 53, 66,
 155, 177, 182, 194–196, 210–
 212, 220, 230, 233, 236, 239
initial resistance to, 59–60
Kelvin and, 38
laws of, 5–6 (*see also* First law of
 thermodynamics; Second law
 of thermodynamics)
meaning of word, 5–6
moving, 108–110
opposition to evolution and, 42
paradox of heat and, 9–10, 25,
 41–44
politics and, 79–82, 85–86, 93,
 96–100, 116
popularization of, 38–39
progressivism and, 41–42
progress of energy and, 49–53
protean quality of, 113–114,
 167
puns and, 23, 115–116, 124,
 128, 214, 280n6, 292n50,
 302n78
scale and, 133–138
spontaneous combustion and,
 190–192
Thomson and, 38
transforming language of,
 114–118
Turner and, 13–14
universal pessimism and, 41–42
ThermoPoetics, 259. *See also*
 Literature
 concept of term, 35
 defining, 35, 72
 engines and, 189, 192

metaphor and, 258 (*see also*
  Metaphor)
progressive approach and,
  225–226
social context and, 31
survival of the fittest and, 112
transformative statements and,
  113
Victorians and, 225 (*see also*
  Victorians)
word work of, 153
Thomson, James, 74, 81
Thomson, William (Lord Kelvin),
  5, 23–26, 29, 125, 187
coining of term
  "thermodynamics" and, 38
conservation of energy and,
  38
counterentropic transformations
  and, 183
Darwinism and, 41–42
death of sun and, 45
dissipation and, 142–143, 237
entropy and, 187
force and, 126, 139–140
God and, 58, 140–141
grand unified theories and, 71–
  74, 90–91, 95, 100
heat loss and, 60–62
initial resistance to
  thermodynamic theory and, 59
nature and, 62
perfect theory of heat and, 60
progressivism and, 225–226
thermodynamics and, 38
waste and, 237

Time Machine, The (Wells), 45,
  225, 227, 229
Transfiguration, 140
Tropes, 40–41, 52, 152, 159–161
Tulkinghorn, 217, 219–221, 227–
  228, 230, 237
Turner, J. M. W., 13
Turveydrop, 196–198
"Two Cultures and the Scientific
  Revolution, The" (Snow),
  22–23
Tyler, Wat, 214, 293n59
Tyndall, John, 26, 71, 91, 125
Belfast Address and, 282n34
energy and, 28
force and, 95, 127
God and, 142
grand unified theories and, 95
heat engines and, 39, 88
media portrayal of, 282n34
Spencer and, 18, 151, 157–158
Stewart and, 282n34
Tait and, 280n11
Tennyson and, 39, 47, 271n56

Universe, 91
basic forces of, 71–72
as closed system, 42–43
conservation of energy and, 6–7
counterentropic transformations
  and, 180–183
entropy and, 6, 192–193 (*see
  also* Entropy)
free will and, 134
grand unified theories and,
  71–112

Universe (*cont.*)
  mechanistic approach to, 13–14,
    187
  moral laws and, 162
  spiritual, 91
  unseen, 90–91, 126–128, 141,
    238
  as variety of actors, 190
*Unseen Universe or Physical
  Speculations on a
  Future State, The* (Tait and
  Stewart), 90–91, 126–128, 141,
  238

Vane, James, 233
Vane, Sybil, 233
Van Helsing, 238, 246, 248,
  251–253, 257
Vectors, 115–121
Vestiges of Creation (Chambers),
  36
Victoria, Queen of England, 34
Victorians, 260. *See also specific
  individual*
  class relations and, 137–139
  control of entropy and,
    301n65
  death of sun and, 23–24, 28–29,
    44–49, 53–57
  energy and, 5, 26 (*see also*
    Energy)
  engine design and, 11 (*see also*
    Engines)
  interdisciplinarity and, 19
  modern cultural similarities to,
    23–24

  motion concept and, 71–73, 77,
    83, 87–92
  positivism and, 49, 72
  Romanticism and, 35 (*see also*
    Romanticism)
  scientific affect and, 15, 24, 29,
    41–44, 49, 62, 64, 155
  sex and, 25, 62–63, 75, 248–252,
    299nn41, 53
  textual fabric of, 25
  university system of, 26
"Vision, A: *Of a Wrangler, of a
  University, of Pedantry, and of
  Philosophy*" (Maxwell), 120
Vis Viva Controversy, 124–125
Vril-ya, 30, 95, 116, 123, 147,
  152, 154, 235
  American empire and, 277n65
  diffusion and, 87
  energy unification and, 75–76,
    79–82
  engines and, 81
  equilibrium and, 82–86, 88,
    104–105, 110–112
  first principles and, 76
  government of, 79–82, 85–86,
    93, 96–100
  living force and, 88–93
  progress and, 100–101, 104
  rest and, 105–110
  Zee and, 78–80, 88–90, 97–98,
    111

Waste, 226
  candles and, 193–194, 206–207
  Darwin and, 29–30

Dickens and, 173–181, 194, 198, 201–206, 216
entropy and, 8–9
evolution and, 29–30
grand unified theories and, 106–107
heat loss and, 60–62
heat sink and, 43
nature and, 15, 29–30
second law of thermodynamics and, 44–49
Spencer and, 78, 78–79
Stevenson and, 228
Stoker and, 235
Tennyson and, 39, 43–48, 57, 61–66, 78–79
Thomson and, 237
Watt, James, 126, 189, 211–214
Wells, H. G., 27, 45–46, 193, 227, 229, 296n8
Westenra, Lucy, 249
Wilde, Oscar, 227, 229–234, 243
Wilkinson, Harry, 7
Work, 5, 8–9, 257, 293n58
as converted energy, 203–204
Dickens and, 152, 155, 157, 165, 167, 172–183, 187, 189, 197–223
dissipation and, 226–228 (*see also* Dissipation)
domestic engines and, 195, 205, 207–212
engine design and, 11
equation for, 203
force and, 122, 131–132, 136, 148
friction and, 122

grand unified theories and, 71, 73, 86, 96, 106, 112
heat and, 242 (*see also* Heat)
Stoker and, 235, 240–241
Tennyson and, 43, 47, 56, 66
Wilde and, 231
Wright, Joseph, 14

X-rays, 254–255

Young, Thomas, 37, 39

Zee, 78–80, 88–90, 97–98, 111